Science and the Spiritual Quest

Addressing fundamental questions about life, death and the universe, *Science and the Spiritual Quest* examines the ways in which scientists negotiate the complex frontiers between their scientific and religious beliefs.

Distinguished cosmologists, physicists, biologists and computer scientists of different faiths explore the connection between the domain of science and the realms of ethics, spirituality and the divine. Through essays and frank interviews, they offer honest, stimulating, and often intensely personal thoughts about life's most impenetrable mysteries.

This unique volume presents radical new approaches to the religion/science debate and highlights the continued importance of the "spiritual quest" in a world transformed by the developments of science.

W. Mark Richardson is Professor of Theology at the General Theological Seminary in New York City. **Robert John Russell** is the founder and director of the Center for Theology and the Natural Sciences, and Professor of Theology and Science in Residence at the Graduate Theological Union, Berkeley. **Philip Clayton** is Professor and Chair of the Department of Philosophy at California State University (Sonoma). **Kirk Wegter-McNelly** is a doctoral candidate at the Graduate Theological Union, Berkeley.

Science and the Spiritual Quest

New essays by leading scientists

**Edited by W. Mark Richardson,
Robert John Russell, Philip Clayton
and Kirk Wegter-McNelly**

London and New York

First published 2002
by Routledge
11 New Fetter Lane, London EC4P 4EE

Simultaneously published in the USA and Canada
by Routledge
29 West 35th Street, New York, NY 10001

Routledge is an imprint of the Taylor & Francis Group

© 2002 W. Mark Richardson, Robert John Russell, Philip Clayton
and Kirk Wegter-McNelly

Typeset in Times by Taylor & Francis Books Ltd
Printed and bound in Great Britain by The Cromwell Press,
Trowbridge, Wiltshire

British Library Cataloguing in Publication Data
A catalogue record for this book is available from the British Library

Library of Congress Cataloging in Publication Data
A catalog record for this book has been requested

ISBN 0–415–25766–2 (hbk)
ISBN 0–415–25767–0 (pbk)

Contents

Some concluding reflections **253**

PHILIP CLAYTON

Preface

Over the course of three years scientists from various fields, drawn from the highest echelons of the academy, met with one another to explore connections between science and spirituality in their own lives. In this book, they share the fruits of their investigation into the compelling and often tangled relationship between science and spirit, two paths which people across the globe continue to follow in their search for greater understanding. These scientists take the spiritual quest very seriously, and their efforts have produced genuine insights.

The essays in this book reflect diverse scientific perspectives: biologists wonder about where life begins and ends; physicists meditate on the relationship between natural beauty and the elusive wisdom of spiritual insight; cosmologists attempt to wrap their minds around the unfathomable reaches of our vast universe; and computer scientists marvel at the power of the intangible stuff we call information.

The essays also reflect a rich assortment of religious backgrounds: a Quaker tests the limits of scientific investigation; a Muslim probes the call of the Prophet to seek knowledge from birth to death; a Catholic reflects on the significance of an evolving universe for an understanding of God's presence in the world; a Jew meditates on the evils perpetrated by humanity as well as those seemingly committed by God; a Presbyterian mulls over the nature and role of authority in science and religion; an Anglican contemplates the balance of chance and necessity in the world; a Hindu praises the unity that lies beneath the variety of our experiences; and an agnostic ruminates on the transcendent aspects of human experience.

Often, these days, it is seen as risky business to mix science and spirituality. In some circles, talking about spirituality can easily place one's hard-won professional image in jeopardy. Sincere personal faith, after all, implies a lack of intellectual honesty and rigor – at least according to conventional wisdom. In light of the suspicion that God-talk often raises, these accomplished scientists have taken a bold step in going beyond their areas of expertise to address such a difficult and controverted topic.

Each of them has written numerous scientific articles and books, and several have written previously on spiritual and religious themes. Bruno

Guiderdoni, in addition to being an astrophysicist, is an expert on medieval Islamic theology and produces a French television show on Islam. Arthur Peacocke, a biochemist and Anglican priest, is highly regarded for his bold reconfiguration of traditional Christian themes in light of science. Michael Arbib, who describes himself as an "atheist theologian," has delivered the prestigious Gifford Lectures in Natural Theology. And Michael Ruse, a well-known and ardent defender of evolutionary theory, has recently written in his inimitable style on the prospects of being a Darwinian and a believer. The desire of these scientists to engage in interdisciplinary writing of this nature reflects a conviction that each of us has an obligation not only to advance our own field, but to contribute to the cultural conversation on matters that extend beyond the confines of the academy.

Although these authors have obtained wide recognition for their scholarly achievements, the essays written for this volume are accessible and crafted with the non-scientist in mind. Some are more difficult than others, but all are worth the effort. Several cover ground that will be familiar to anyone who follows current events, while others delve into exciting, unfamiliar topics. Each author draws upon mastery of his or her scientific discipline to convey a sense of the struggle and insight, the disappointment and accomplishment, that come with a life devoted to seeking answers to difficult questions.

The book is greatly enhanced by the interviews which precede each essay and help to situate a scientist's thought in the context of his or her life's journey. Gleaned from hundreds of pages of transcripts resulting from intensive give-and-take conversations with Gordy Slack and Philip Clayton, these interviews provide the kind of personal insights that rarely find their way into more formal compositions. Through the interviews and essays, the voices of these scientists emerge as remarkably bilingual in the languages of science and faith and far richer for holding multiple points of view in conversation.

Some scientists fashion themselves uniquely positioned to be the meaning-makers of our age. Unfortunately, such self-anointed experts often have no interest in or knowledge of the subtlety of spiritual and religious matters or traditions. The scientists in this book are clearly different. Each brings to the spiritual quest a refreshing sense of humility, as well as an eagerness to ask hard questions – about the meaning and purpose of our lives and our place in the world, about suffering and death, about physical matter and what matters most, about the prospects for our planet's well-being and the future of the cosmos, and yes, about God. While none would accept the title of high priest, each is familiar with one or more of the world's major religions, some having grown up in a religious environment, others having embarked upon their spiritual journey later in life. Each shows a degree of sensitivity to the kind of spiritual questions that propel us beyond the capacities and limits of science.

The essays presented here are particularly helpful in identifying possible topics and modes of conversation, for none assumes the mistaken and outdated view in which science and religion are locked in unending, fruitless conflict. This book is a powerful example of the benefits that come from genuine interaction between science and spirit, especially when such interaction draws on the rich resources of the world's great religious traditions.

Perhaps most strikingly, these scientists articulate a sense that letting go of science's claim to uniquely reveal the world as it *really is* yields not so much the loss of stability or certitude – which, at any rate, were dangerous illusions, whether conjured by science or religion – as the discovery of a more authentic freedom which, in the long run, will better equip science to deepen its contribution to the human quest for understanding.

As a society, we have come to derive satisfaction from focusing our resources on those problems we think will likely yield under pressure. As participants in an increasingly technologically oriented and affected world, we pride ourselves on being able to provide ourselves and others with solutions. All the more remarkable, then, that these scientists have chosen to venture beyond the comfort and familiarity of their desks and laboratories to reflect with one another on questions which resist the analytical tools they are so adept at wielding, questions which hold our interest not because we can respond with definitive answers, but because asking them makes a difference in what we believe and how we live.

Kirk Wegter-McNelly

Introduction

W. Mark Richardson

What is this volume about?

Scientists are accustomed to experimentation, but this book is the result of a very unusual experiment indeed. This volume presents leading figures in the sciences, widely published in their respective fields, who now venture to write about the interdisciplinary ground between their science and matters that are *spiritual*. The philosopher and theologian, Paul Tillich, named this the dimension of "ultimate concern": the source of meaning that structures and interprets all aspects of one's experience.

This writing was inspired by a project named Science and the Spiritual Quest (SSQ). In 1997, sixty leading scientists from around the world met in SSQ workshops in Berkeley, California to discuss together the relationship between theory in their science and major themes from some of the world's great spiritual traditions. It is not so rare for scientists to have been shaped and influenced by one religious tradition or another, but it is very unusual, perhaps unprecedented, for scientists to meet with colleagues from different traditions, to reflect on the relationship between these aspects of their lives; and to consider the areas of consonance and dissonance they find at the points of interaction between science and religion.

These essays were born from the discussions and interactions that occurred at the SSQ meetings. There are physicists and astrophysicists, biologists and computer scientists, all representing major themes which emerge at this interface between two of the most powerful forces in human cultures today.

The value of this dialogue

Should we take such conversation between science and spirituality to be a cultural breakthrough, or a step backward from twentieth-century ideals? Throughout much of the twentieth century, intellectual culture preserved a radical independence of science (as publicly accessible objective knowledge) from religion (as subjective ideas based on personal preferences derived from the contingencies of one's background). Under this perspective, religion, spiritual practice and even metaphysics were irrelevant to the pursuit of knowledge. Others in this era were more explicitly hostile to traditional ideas of Western religion, seeing them as simply wrongheaded and in conflict with the present scientific worldview. Ironically, this latter view at least affords matters of religious scale enough cognitive value to merit the status of "in conflict." In neither model, however, is anything informative thought to emerge at the nexus of science and religion.

Why are these scientists writing about these matters? In many cases the essays constitute a tacit rejection of the dogma of "separation," the idea mentioned above, that science and religion have no grounds on which conversation between them can occur. Most of these sixteen scientists see the intersections between intellectual frameworks of interpretation produced by the spiritual quest for meaning and the theories and data that inform but cannot create a meaningful worldview.

We often marvel at the success and power of contemporary science. In countless ways it is transforming Western civilization. But rarely do we note that science of today originates in the cultures of Western monotheism – Christianity, Judaism and Islam. This connection seems distant, under the apparent strain between religion and science in our day. But if we probe beyond appearances and biases, as these scientists have done, there are hints of science's context within a larger framework of the human quest for meaning and purpose. Is religion's pursuit of ultimate concerns insinuated in the scientists' pursuit of truth? Will common features in the practices, experiences and disciplines of science and of religion be discovered and acknowledged?

Most of the scientists represented in this volume recognize the influence upon them of one or another of the monotheistic traditions. They wish to explore how these formative experiences have influenced their professional work. Conversely, they explore how the new perspectives gained through the sciences have affected their understanding of the great religious themes about God: the nature of the human person as moral and spiritual agent, the purpose and meaning of the universe. Some of these scientists are very tentative about the way they express their views on these topics, others are more bold. In all, one senses subtle and complex factors in the relationship between scientific and religious pursuits of truth.

Others represented in this volume do not acknowledge faith in a transcendent reality, but recognize the human quest for meaning and are in pursuit of a coherent moral vision, consonant with the scientific worldview, that will capture the motivational core of human beings.

Notable themes and features of these essays

There are recurring themes running through the essays.

- A noticeable wrestling with the *status* of spiritual insight or "truth," and the wisdom of tradition: it resonates deeply in human experience, yet lacks the "test-ability" and exactness we demand in the sciences.

- The theme of religion's *moral* center surfaces often, with several scientists finding the heart of spirituality in its contribution to moral issues and policy decisions related to scientific research and application.
- A number of the scientists remark on the limits of science: what it does, it does well, because it limits itself to certain kinds of measurable properties of objects and states of affairs. So how can it touch upon questions of such breadth as those posed by spiritual traditions?
- Others wish for the major spiritual traditions to recover an open and inquisitive spirit (and through this recover their vitality). They worry that religion is locked in dogmas that are kept out of contact with living experience and has lost its power to interpret experience.

Ironically, one burning issue in popular culture that drives a wedge between religion and science – the apparent conflict between evolutionary theory and theological ideas about "creation" – seems not to motivate these scientists at all. This, of course, is the most passion-filled nexus of science and values in American culture today at the level of public school education and other areas of popular culture. No one seemed burdened on a spiritual level by Darwinian biology and its implications regarding the human person. Those that consider the question at all recognize that evolution is open to various interpretations and does not have devastating effects on their spiritual self-understanding.

Consonances and dissonances between different domains and modes of knowledge and experience come and go as a result of this dynamic flux in all spheres of knowledge. This works against the expectation that science will, at all times, be complementary with long-held perspectives rooted in religious tradition. While religious traditions may, and indeed must, change themselves, the pace of such change depends on many factors and will not always be determined by changes in the sciences.

Readers will be struck in these essays by how the classic questions of every age remain central in this one, regardless of scientific advancement. Concern for:

- the reality of pain and suffering against the backdrop of faith in God's goodness;
- balancing nature's unremitting regularity with the intuition of freedom, and discerning the place of God in this;
- making sense of belief in the *purpose* of life and the universe in the face of the ambiguity of natural evidence, and science's methodological resistance to the search for meaning;
- finding the meeting place between scientific pursuits and the moral responsibility of the scientific community within the larger culture.

What is noteworthy about the attitude of these scientists on these issues, and impressive in their conversations, is their seriousness and reverence with respect to classic religious issues, without dogmatism. The open, exploratory spirit of science – so fundamental to its success – is carried over to these questions about cosmic and human origins and destiny, and about the source of being and object of our worship. There is no false certitude, and no stubborn refusal to consider the evidence from any source of human experience. We see here a model for something that may be informative to broader world religious dialogue: stepping back from our respective certitudes and adopting a posture of deep listening; receiving the voice of the "other" so that it can inform and expand our own long-held convictions.

The diversity of opinion

The insight that arises most clearly in this collection is that there is no single answer to the questions posed at the juncture of science and religion. Although there are agreed upon elements in the practice, method and theoretical outlook of each scientific discipline, the practitioners themselves are persons with many and varied influences on their lives. As a consequence, each appreciates the mutual influence of science and religion in his or her own unique way: some acknowledging a high degree of integration, others complementary relations but real differences in the modes of knowing and objects of knowledge, still others are uncertain that the two domains of their lives interact extensively at all.

Indeed the variety of possible understandings among people who share so much in common as scientists is part of the story told by this book. One's first impression is of a rich, almost startling diversity; clearly the sixteen scientists who have spoken and written in these pages do not reflect a "party line." Instead, they represent a vast array of positions running from strict atheism through agnosticism to the deepest forms of devotion to God. Some draw very tight lines of connection between their science and their religious belief; others are cautious about even the most limited of connections.

How are we to make sense of diversity such as this? If science depends upon repeatable results, what does it mean that a vast range of interpretations pop up once we interpret science on this level of human quests for meaning and spiritual fulfillment?

When systematic inquiry in any given field of natural knowledge takes us to the edge of that field, it often leads the inquirer to apprehensions beyond the discipline itself. A well-known contemporary example is the breakdown in the equations when one reaches singularity in classical Big Bang theory. Once we have pushed back as far we can toward the singularity, we finally face the question of cosmic origins and the fundamental question "why is there something and not nothing?" which has occupied human imaginings since the dawn of civilization.

These boundaries demonstrate the inherent limitations of the scientific method. Nevertheless, there *is* a relationship: what physics can tell us affects the relative viability of the various metaphysical options. Conversely, the metaphysical commitments dispose us toward the interpretations we make of the available scientific evidence. This is a peculiar situation: no necessary relationship, but real interdependence. For example, "chance" in nature signaled meaninglessness for the philosopher of biology, Jacques Monod, but it indicates insight into the creative activity of God for biochemist and theologian, Arthur Peacocke. They each have access to the same science. The science as such cannot arbitrate their differences in interpretation, but the philosophical commitment only survives if it can make sense of the known data. This is a situation somewhere between necessary entailment and complete independence.

All too frequently the large-scale commitments that guide conversation and writing at the nexus of science and religion go unstated. Consider the diversity of views in the following list:

- Paul Davies writing about the *"mind of God"*;
- E.O. Wilson interpreting religion through the lens of socio-biological explanation;
- John Polkinghorne's and Arthur Peacocke's theological revisions based on contemporary science;
- eco-feminists using feminist principles to reconstruct a contemporary cosmology;
- Brian Swimme's and Thomas Berry's mythic-like narratives of the universe.[1]

All of these are attempts to pose worlds of meaning within a contemporary scientific perspective. The range of viewpoints in this list alone is staggering. How do we sort it out?

Understanding the diversity: a typology

A typology might be useful here for understanding the ways that many today are making the links between science and religion. One way to categorize contemporary explorations at the nexus of science and religion is to use three common motifs in nineteenth-century intellectual culture.

- There are those who adopt the Enlightenment quest for universal and ahistorical principles of reason by which to speculate on matters of ultimate concern. We will refer to this as the *rationalist-speculative* type.
- Others represent something analogous to the Romantic movement's reaction to this: a drive to restore a sense of cultural *telos*, to elicit the affective dimension in the understanding of one's place in the universe, and to appreciate a fundamental unity of consciousness and personhood with the whole of the cosmos and nature. We will name this the *affective-holistic* type.

- Still others place stress on the cognitive dimension of religious belief, applying critical historical and philosophical bridges to connect what appear to be very different conceptual systems of science and religion. Here there is recognition of interaction between science and religion, but also distinction between them and their institutional histories. This last group will be referred to as the *critical-historical* type.

I will concentrate on issues of *language, attitude* and *method*, to characterize the types and to draw out the distinctions.

Rationalist-speculative approaches

The rationalist-speculative type is not invested in a rigorously developed link between the scientific view of nature and historic religious traditions, especially Western theistic traditions, which stress the personal source of all existence. Consider, for example, Andrei Linde, SSQ scientist and Professor of Physics at Stanford University:

> from the beginning this understanding of God as somebody sitting and creating the universe was from, well, bad movies. There are some primitive ideas in the minds of people until they have found something better. The question is, what is the deepest level of ideas about God that may survive after science investigates things?
>
> (interview)

The rationalist begins within the rigors of the sciences, and from there engages in the speculative task of building inferences about principles of ultimacy from this rational and empirical grounding in the sciences. There is optimism about uncovering universal and timeless principles of scientific rationality that, when followed, will lead to understanding the basic laws of the universe. From these we will be able to speculate about matters of meaning and value we attach to our experience of the universe, informed by human judgment and knowledge as well as by physical cosmology.[2]

This type does not invite much self-involvement or affective attachment in this process. As Paul Davies states, "There is a long and respectable history of attempts to confront such issues by rational and dispassionate analysis."[3] The suggestion here is that it is scientific rigor that leads to knowledge of the world, which can then be generalized in meaningful ways.

The movement is in a single direction: beginning within the standards and methods of one's science and moving toward its boundaries, at which point one looks for connections, or patterns that can be extended as inferred from the science, and applied as ultimate principles lying beyond science *per se*. This becomes the basis for all cognitive claims.

Exemplars of this type may be suspicious of any ideas or assumptions derived from ordinary human experience. The potential for delusion in common-sense experience can make it an unsatisfactory route to knowledge. In contrast, it is thought that the natural sciences, especially physics, lead to results which may be counter-intuitive. Knowing the "cosmic code" through science and following the assumptions of common-sense experience can be at odds. If this is the case, then those who are trained in the intellectual habits and procedures of physics are in a privileged position to provide "true" principles of the universe. Again, Andrei Linde:

> When you look around yourself, there are some solid, well-established facts of your life. This is quite satisfactory for 99% of the population. But then somebody tells you that space is curved, and that parallel lines might intercept somewhere, and that the universe may be closed on itself, and that particles are not particles but in some sense waves. These things are so much against the intuition of your everyday life. Nevertheless, even though these things seem like a miracle and are absolutely anti-intuitive, they are solidly grounded in truth and in experimental evidence.
>
> (full interview)

If principles of ultimacy are going to be shaped directly out of science, and if we forecast science's continued success, then according to some exemplars of this type, we may be optimistic about eventually pushing back the boundaries of mystery, and resolving some classical questions thought to lie beyond science. In discussions in a workshop for physicists, the question arose over the distinction between *puzzles* (i.e., problems to be solved) and *mysteries* (i.e., those things that are not known via science, and thought to be in principle unknowable). "Does physics recognize this difference?" we asked. In reply, Nobel physicist Charles Townes, and SSQ participant, rejected the distinction. His claim was that if science were to pre-determine what cannot be known, progress would end. Resort to mystery closes down the inquiring spirit of science. Many have argued, for example, that the question of human consciousness and freedom lies outside the full reach of science. SSQ scientist, Geoffrey Chew, retired Professor of Theoretical Physics at the University of California, Berkeley, and inventor of bootstrap theory, states, "I deeply believe that absolute truth is unobtainable." Nevertheless, in the spirit of this unrestricted quest within the sciences, he goes further to state that many phenomena thought to be mysteries beyond science must not go unexplored scientifically: "We have to know what human self-consciousness means. Science cannot ignore the phenomena of 'soul'" (full interview).

Given the central role of science in this type, the language it employs at the science–religion interface can be austere and impersonal. The goal is to remove all elements from the interdisciplinary engagement which would individualize or personalize knowledge and judgment. At the juncture linking science to ultimate concerns, exemplars of the intellectual type either turn silent and retreat into stances of personal commitment (that are regarded then as personal preferences); or they speculate about the other side of the boundary by way of inference from science itself, staying close to the conceptuality of science in doing so. Paul Davies, for example, when exploring frontiers of meaning, highlights the combined features of *order* and *contingency* evident in the universe, and finds the insinuation of intelligence underlying the universe. Freeman Dyson, impressed by the scientific observation of *diversity* on so many fronts, develops the metaphysical value of this feature.[4]

Although this cosmological picture may engender awe and wonder, there is little or no resource for capturing the aspect of self-involvement, and little need to remain in tension with the rich symbolic power of religious traditions on such matters. The problem posed by this type is whether the God derived from natural theology (if, indeed, "God" is a term seriously employed) is a God that can be worshiped, or whether the principles of the (C)osmology thus derived can inspire human passion and direction.

Affective-holistic approaches

Those who have the affective and holistic orientation toward the relating of science and spirituality are driven by the quest for unity of one's consciousness and being with the cosmic whole. To achieve this goal, they see the need to recover a cosmology that derives from the most integrative picture we have from contemporary sciences, that we then invest with a worldview which underscores this thesis of cosmic wholeness and synthesis.

SSQ scientist, Pauline Rudd, molecular biologist at the glycobiology labs of the University of Oxford, and member of the Church of England, states:[5]

> As Wordsworth knew, there is a kind of intuitive wonder that we experience when we look at the natural world and see a glorious sunset or a leopard running across the plains ... eventually this type of intellectual wonder can give way to the intuitive, for the mechanisms can become so well studied and understood that the distance between the observer and the observed is somehow eclipsed and the scientist understands the sunset from the inside. Perhaps this is like those in some spiritualist traditions who claim to be transformed into animals or trees, or to be able to predict the weather. I, for example, would like to be able to empathize with the molecules I work with.

An SSQ scientist represented in this volume, George Sudarshan, a physicist at the University of Texas, expresses the unity of life and knowledge, from the perspective of his grounding in the spiritual life of Indian culture:

> My tradition affirms that any spiritual search, whether academic or not, is bound to lead to God. Within Hinduism, there is nothing which is not sacred. God is not an isolated event, something separate from the

universe. God is the universe. ... Yes, God is more than the universe, but He is the universe.

(essay)

The expression of this integrative vision is often in narrative form and in highly symbolic language. This linguistic strategy reflects the unity sought after, for the story is a synthesis – a bringing to oneness of cognitive, aesthetic, affective and moral dimensions of knowing into a single language.

SSQ scientist, Joel Primack, theoretical physicist and cosmologist at UC Santa Cruz, and discoverer of cold dark matter, states:[6]

We may see in the first decades of the Twenty-first Century the emergence of a new [scientifically inspired] universe picture that can be globally acceptable, and with this and the contributions of image-making writers, artists and spiritual visionaries, it is possible that the painful centuries-long hiatus in human connection with the universe will end.

This quest evokes a connection of the self with a larger whole. It is recognized that cognitive, aesthetic and moral dimensions must be drawn upon to make this connection. Consider the following example from Brian Swime and Thomas Berry's *The Universe Story*. They use the mythic name Tiamat, the ancient Near-Eastern god, to identify supernovas, the stellar deathly violence that brings new diversity and eventually rich communion. By personalizing a natural force they evoke intentionality in the rendering of the cosmic story, making the bond between personal existence and the whole of reality. Similarly, the same authors state, "the eye that objectively searches the Milky Way is the very eye shaped by the Milky Way in search of its own inner depths."[7]

The language is personal, and personally engaging, to reinforce this fundamental unity of consciousness with the cosmic whole. Says Pauline Rudd:

We need to develop a language to express these intuitive ideas ... and it is symbols – used by science, religion, music and painting – that bridge between earth and heaven; between intuitive and rational understanding.

As with the rationalist-speculative type, there is no clear motivation to preserve the institutional histories of Western theisms, or for that matter, any tradition at all. Historical traditions are seen as unlikely vessels for holding the spiritual sensibilities of the contemporary, scientific age. The effect is to relax the need to relate, in any rigorous and systematic way, the conceptually distinct systems and traditions of science and religion. The emphasis here falls on synthesis.

According to the affective-holistic approach, history only conveys the movement toward unity that one intuitively finds in the present context, a general movement toward a goal of complexification, or self-creation, or diversity within unity. This cannot be done through the austerity of scientific language alone. Consider, again, the words of Swimme and Berry:

> Any cosmology whose language can be completely understood by using one of the standard dictionaries belongs to a former era ... an encounter with the new cosmology is a demanding task, requiring a creative response. ... Human language and ultimately human consciousness need to be transformed to understand in any significant way what is intended.[8]

The pressures of reality, according to these authors, drive the creative and questing spirit to push against the boundaries of language, and to be an insurrectionist and bold form of expression.

In conclusion, the strength of this type is that it elicits a passion and self-investment in this dialogue between science and spirituality. This is accomplished through the power of synthesis, drawing upon the affective dimension of humanity's quest that is missing in the rationalist-speculative model. This sometimes comes at the expense of critical examination of the various *cosmic* stories. Whose rendering of the "new story of the cosmos" are we to believe, and on what grounds? Does the sense of unity achieved come at the cost of glossing over the complete scientific picture of nature – one that includes not only its beauty but also brutality? This absence, however, is not something intrinsic to this type and is rather a tendency to be guarded against.

This model, more sparsely represented in this volume, was nevertheless present in the outlook of some of the scientists in the SSQ project.

Critical-historical approach

What concerns those within the critical-historical type is the way the historical tradition functions in the conceiving and conditioning of the present understanding of the God–nature relationship. There is sensitivity to the cumulative and progressive relation between conceptions of God and nature. New knowledge from the sciences reveals new integrations with religious tradition, its mysteries, and its positive conceptions of God. It is a recognition of the way the past is embedded in present communities of inquiry. For this reason, those who are committed to traditional theistic traditions are attracted to this outlook and become its exemplars.

As Arthur Peacocke, a biochemist, Anglican priest and SSQ scientist, states it, the task at this interface of contemporary science and spirituality is "to rethink our religious conceptualizations in light of the perspective on the world afforded by the sciences."[9] This involves not only the mixing of historical frameworks (ancient creeds and historical texts) but philosophical and conceptual ones as well. This is *not* a burden for the other approaches because they are less tied to the formalized and integrated belief structure of Christian theism.

Peacocke's way of addressing the philosophical bridge is through the development of an epistemological hierarchy that allows us to see continuity and differences at various levels of natural organization. We find a hierarchy of complexity from sub-atomic, atomic, molecular, cellular levels and so forth, all the way to the most complex levels of organization in the universe. Theology is the most integrative of disciplines concerned with meaning and value. It is constrained from below (by psychology, social theory and so forth), just as theory in molecular biology is constrained by physics (we won't tolerate explanations in biology that violate laws of physics).

In spite of constraint, each new level of complexity requires explanatory concepts at its own level representing functional properties at work on that level. Through developing disciplinary maps such as this, Peacocke attempts to show how science and religious themes can be related and he sees this kind of integration as a necessary step toward any realization of theology's role in public conversation.

Another SSQ scientist, Martinez J. Hewlett, Professor of Molecular and Cellular Biology at the University of Arizona, follows in the historical-critical path, and echoes Peacocke's methodological concerns. Hewlett sees that careful attention to methodological and historical issues, and to the nature of the object of discourse, will help us see that apparent conflicts between science and religion are often no conflicts at all. In his nuanced essay on the nature of reductionism, its correct use in science, and its potentially abusive use in extra-scientific contexts, Hewlett states:

> Investigations of material reality should be informed by, not at odds with, investigations of spiritual reality, and vice versa. There is no need for scientists to overlook what is perfectly obvious in everyday experience, even if it cannot be measured by instruments.
>
> (essay)

With an interest in the same question of reductionism, SSQ scientist, Robert B. Griffiths, a physicist at Carnegie-Mellon University, contemplates the relation between reductionism in physics and discourse about the scale of reality central to personal and religious experience. He suggests that a careful use of reductionism in the sciences is, in the end, an ally of theological thinking. Pressing reductionism to its limits will make our understanding of human agency, for example, better informed, and will expose the limits of physical description in our accounting for human action and moral responsibility. His essay in this volume is a clear statement of the importance of both distinguishing and respecting different levels of discourse important to human enterprises, some springing from our sciences and others from personal and religious scales of experience.

Here I would like to mention the thinking of Bob Russell on this issue. Russell, the Founder of the Center for Theology and the Natural Sciences, in Berkeley, California, considers the relationship between classical Big Bang cosmology and the monotheistic doctrine of creation as one of consonance. He stresses the fact that, in trying to locate the consonance between theology and science, one must acknowledge both similarity and difference in their respective conceptual systems. This reminds us that any effort to show relations between science and the spiritual quest requires considerable care. One will find points of relationship, but one should not expect to find them on the surface of our language in the compared systems. Consider the following diagram:

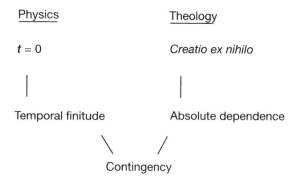

Notice that theories in physics and theological beliefs represent fundamentally different conceptual domains. There is no immediate connection between "$t = 0$" and "*Creatio ex nihilo*" in the diagram. Nevertheless, they may share common metaphysical roots. The relation is not direct and must be mediated through the steps shown in the diagram under each tradition, but each can be traced back to a shared fundamental notion of *contingency*.

Coupled with the appreciation for the complementary relationship between the institutions of science and religion, the historical type also is sensitive to the distinct functions and tasks of science, on the one hand, and the worlds of meaning we typically associate with our religious traditions, on the other. Each has its own institutional history, its own functions and language, so that the relationship between them can be complex. The challenge is in finding the appropriate way to mediate the difference so that true connections between them can be located.

SSQ scientist, Bruno Guiderdoni, astrophysicist at the Institute of Astrophysics in Paris, believes his Sufi Muslim tradition provides a motivation for the pursuit of all aspects of knowledge, including science, and instills a moral attitude about this pursuit:

> I think they [religion and science] are complementary. Islam strongly emphasizes the importance of knowledge in life in general and in religious life in particular. ... There is restriction in Islamic tradition because the knowledge that must be looked for is useful knowledge, that is, useful to mankind in general. That means that the pursuit of knowledge is not separated from ethical values. ... As a Muslim, I feel very comfortable in my scientific activity because I can interpret my research work as the pursuit of knowledge of this world, as the exploration of the richness and beauty of God's Creation.
>
> (full interview)

The apparent conservatism of the historical approach is traceable to both religious and philosophical convictions. In the present postmodern context we have learned that there is no privileged context, no jumping out of our histories in the pursuit of truth and first principles. The emphasis on the historical context of concepts and ideas does not entail complete relativism – there is still a possibility for inter-subjective conversation, even inter-subjective judgment about explanatory schemes. But the foundations, as they are understood in the rationalist model, are looked on with some suspicion; there is more recognition of finitude, of limits.

Western theisms are beginning to appreciate the social dimension of their beliefs and practices. Awareness of one's connection to communities of belief through time leads the historically conscious to an appreciation of the traditional sources of wisdom and truth. Again Guiderdoni:

> In spite of these [scientific] discoveries, the important thing is not there; the important thing is man's spiritual realization because we need more knowledge than science is able to give us. ... If we want to address the question of origins, we have different kinds of answers. We have the answer given by modern cosmology, that is the way of projecting the question of the origins onto the plane of scientific method with concepts such as space, time, and matter. We also have the spiritual and mystical approach.
>
> (full interview)

The critical-historical type recognizes that there is a gap, at the very least a conceptual gap, between the language of science and the language of our religious traditions. This gap needs a philosophical bridge if there is to be any hope of integration. It is understood that science may originate within Western monotheistic cultures and share common ideas regarding the rationality and contingency of the universe, but religious history and scientific history have their own independent trajectories. The task at the interface of science and religion is both philosophical and historical: to find intelligibility in the various domains of knowledge and commitment in which we take part.

SSQ scientist, George Ellis, a physicist at the University of Cape Town, South Africa, and a Quaker, construes the relationship in this way:

> Science attains near-certainty by limiting itself to very specific quantifiable issues, but consequently cannot look at many issues of vital importance to human beings. Theology uses broader classes of data that deal with much wider issues – of major significance to everyday life – where much less certainty is attainable. Both rely on judgment and discernment in their practice and allow a similar attitude of questioning and process of testing.
>
> (from conference presentation)

According to the critical-historical type, there must be sensitivity to the philosophical questions that arise at the interface of religious tradition and contemporary science, questions about:

- *language*, e.g., how do models and metaphors function in science and theology?
- *epistemology*, e.g., along with its affective and moral dimension, does religious language refer to the world; does it have cognitive value?
- *method*, e.g., can theology develop a public methodology that demonstrates explanatory power and a predictive aspect, analogous to the successful methodology of scientific research programs?
- *practices*, e.g., what are the goals and activities of religion and science that constitute their differences?

The difficulty with this approach is: how does one persistently engage in the self-critical stance, as expressed above, without a loss of immediacy in relation to one's spiritual/religious heritage? What effect on the religious attitudes of, for example, awe, praise or sanctity, can we expect from such constant self-critical examinations of religious intelligibility?

These writers offer unique perspectives based on their genius and depth of understanding in their particular fields. A typology cannot by itself exhaust the subtleties and finer points of their respective contributions; moreover, exemplars may at times express impulses associated with more than one type, even if there is a predominant tendency in their work. This typology is simply intended as a useful guide in recognizing motifs that arise at the nexus of science and spirituality.

The SSQ project: the context for these essays

Something more must be said about the larger context in which these interviews occurred. The workshops mentioned above were part of a three-year project, Science and the Spiritual Quest, which involved sixty scientists in research, group discussions and conferences. The workshops occurred in four groups of fifteen, organized around professional disciplinary specialties of physics, cosmology, biology and computer science. Each group met twice, for three days each time, over a period of one year.

Something special occurs when scholars and scientists are encouraged to give utterance to ideas, and to think through the implications of their work *together*, rather than pursue such work in private. The workshops were a time of deepening personal understandings at this nexus, gaining a power of expression on such issues, and discussion of these topics with professional colleagues. This last point is nearly unprecedented in the scientific culture today. Through trust and through open exploratory process, the discussions with colleagues drew out a level of reflection in each person in ways that private reflection could not. The scientists were encouraged to bring a spirit of open, hypothetical inquiry, typical of the process of their scientific work, into discussions involving moral and spiritual topics.

Naturally, there is a deeply autobiographical aspect to this kind of reflection, and this comes through in the tone of these essays, in some cases more than others. The personal context of each scientist, his or her movement from struggle to tentative resolution, and from creative idea to systematic expression – all of this is detected in this genre of writing. One senses the intellectual drive and passion that has made these figures leaders in their fields. They reflect an honesty and openness to criticism and revision in matters of spiritual significance that also characterizes their work as scientists.

The results of this process of workshop discussions and research were presented in a public conference in June, 1998, at the University of California, Berkeley. There, twenty-seven of the participating scientists presented the findings of their research. Never before had such a distinguished group of scientists convened to speak about science and spirituality. As a consequence, over thirty national correspondents and reporters were in attendance at the four-day SSQ conference.

The conference proceedings inspired a *Newsweek* cover story based solely upon the views of several SSQ participants, and an NBC-produced segment about the conference on its weekly *Religion & Ethics Report*.

The editors acknowledge, with deep appreciation, the distinguished work of the Center for Theology and the Natural Sciences (CTNS), which made this project and this book possible. We wish to thank especially Holly Vande Wall for her careful editorial work on the interviews and essays. It was because of CTNS's fifteen-year track record of engaging in cutting-edge research on the interaction of theology and the natural sciences that the John Templeton Foundation generously and confidently funded Science and the Spiritual Quest. Finally, we acknowledge with appreciation the generosity and visionary thinking of the John Templeton Foundation, which has supported this and other very important projects in the field.

Notes

1 Paul Davies, *The Mind of God: The Scientific Basis for a Rational World* (New York: Simon & Schuster, 1992); E.O. Wilson, *Consilience: The Unity of Knowledge* (New York: Knopf, 1998); Arthur Peacocke, *Theology for a Scientific Age: Being and Becoming – Natural, Divine and Human* (Oxford: B. Blackwell, 1990); John Polkinghorne, *Belief in God in an Age of Science* (New Haven, CT: Yale University Press, 1998); Karen J. Warren, ed., *Ecofeminism: Women, Culture, Nature* (Bloomington: Indiana University Press, 1997); Brian Swimme and Thomas Berry, *The Universe Story* (San Francisco: Harper San Francisco, 1994).

2 In a helpful book titled *Before the Beginning*, mathematician and astrophysicist, George Ellis, makes the distinction between (c)osmology and (C)osmology. The former refers to physical cosmology as it is carried out professionally in the sciences by theorists and observationalists. It is a narrower use of the term. The latter, in Ellis's use, refers to matters of meaning and value we attach to our experience of what is the case, that, of course, is informed by our scientific perspectives on the universe and life within it. Whereas many features of physical cosmology may be incontestable relatively speaking, this larger sense of (C)osmology is open because we cannot so easily test unique phenomena, or the meaning and value we draw from our view of ultimate things. I adopt Ellis's idea here and refer to the following questions relating to (C)osmology. See George F.R. Ellis, *Before the Beginning: Cosmology Explained* (New York: Boyars/Bowerdean, 1993).

3 Paul Davies, *The Mind of God: The Scientific Basis for a Rational World* (New York: Simon & Schuster, 1992), p. 20.

4 Freeman J. Dyson, *Infinite in All Directions* (New York: Harper & Row, 1988).

5 Pauline Rudd's essay is not represented in this volume but a more extended view of her understanding at the nexus of science and religion can be found in Routledge's companion volume to this, edited by Richardson and Slack. Although Rudd shows strong leanings toward the affective-holistic position, she shows a stronger appreciation than others in this type for the deep and abiding elements in the historical religious tradition of Anglicanism, from which she comes.

6 Joel Primack's essay also cannot be found in this volume. However, the Routledge companion volume of interviews offers a more complete picture of his treatment of this topic. Primack also shows a tendency to combine features of both the rationalist and affective models, once again indicating the value of viewing typologies only as general guides.

7 Brian Swimme and Thomas Berry, *The Universe Story* (San Francisco: Harper San Francisco, 1994), p. 45.

8 Ibid., p. 24.

9 Arthur Peacocke, *Theology for a Scientific Age: Being and Becoming – Natural, Divine and Human* (Minneapolis: Fortress Press, 1993), p. 4.

1 Jocelyn Bell Burnell

Jocelyn Bell Burnell received her Ph.D. in Radio Astronomy at Cambridge where she was involved in the discovery of pulsars. She is Dean of Science Faculty at the University of Bath and has served as Vice President of the Royal Astronomical Society. She is recipient of the Oppenheimer prize, the Michelson medal, the Tinsley prize and the Magellanic premium, and is an active member of the Religious Society of Friends (Quakers).

Interview by Philip Clayton

PC: I'd like to start on a personal note and ask about your own religious background and the way that was affected by your scientific training.

JB: I was born into the Quaker religion, brought up as a Quaker, and I am a Quaker. I moved to England from Ireland when I was 13 for my secondary schooling and my college education, and I have stayed this side of the water ever since. Quakerism in England is much more liberal and often academic, much less likely to be explicitly Christian, let alone evangelical Christian, and much less likely to be biblical literalist than the Irish Quaker church I grew up in. I've always been significantly involved. Recently I was clerk of the yearly meeting. I suppose the nearest secular interpretation is president of the annual meeting for the Friends in Britain.

PC: Suppose I had met you during your graduate school years, or early in your career, and had said, "How can you take your physics seriously and at the same time be a theist?" How would you have responded?

JB: I suspect much the same way as I would now. My experience of the religious is in other domains of life besides physics. Physics or science is not the totality of life, there are other dimensions. I think some people might feel that I compartmentalize too strongly, but I would argue that I'm seeing a spectrum, a continuum. Starting with what I'm labeling the religious, I'm referring to experience in Quaker meetings for worship, the sense of an immanent God that can be in those gatherings. I'm referring to the promptings or nudgings or leadings that one should move in a particular direction or do a certain thing that one sometimes senses.

   On the science side are the very rigorous, tightly argued, logical processes of testing hypotheses and drawing conclusions, which I think have been compartmentalized too strictly. I think much of what we teach about the scientific method is bunk. Having been involved in a major discovery, I am fully aware that the way discoveries are made and the way conclusions about unexpected things are worked out is not a simple, linear, well-organized process. It is round about, it is to and fro, it's backwards and forwards. I see strong parallels between the way scientific understanding grows and the way one's knowledge of God grows. You have a hypothesis, you test it by experiment. You see what the conclusions are. You revise the hypothesis, and you go round the circle again until you get convergence. In devising hypotheses, there clearly is need for insight, intuition, even dreams. One needs all kinds of faculties for making that leap to devise a new model, a new hypothesis to test. I also suspect that different people will see different things in a model or a hypothesis and may come up with different aspects of it, different questions about it, to be tested by experiment. This is where I suspect cultural differences come in and where there might be a feminist science that is different from the male science.

PC: Are these sorts of subjective factors that you've just mentioned parallel to the inner wisdom or light that a religious person knows?

JB: I think that's true, although I'd be inclined to say that a lot of religion is subjective. For me, faith has to be rationally sound, so there is an academic or intellectual content to it. However, the whole of the meeting for worship runs on intuition and other subjective factors. Although there are very strong parallels between the science and the theology, the balance between the academic and the subjective is rather different in the two, with religion and its practice having much stronger emphasis on the subjective.

PC: Could this be an explanation why so many great scientists have been Quakers, that they are already used to a process of submitting their own inspirations to a broader community, and that's great training for being a scientist?

JB: It could be. I certainly hope that one of the outputs of Quakerism is that one's trained to be rigorously honest. It's not to say that anybody else is dishonest, but because Quakers see everybody as being of equal value, it does mean one's less likely to get an inflated impression of one's own ability and standing. I think that tempering can also lead to honest questioning.

PC: You're a theist, you believe in a spiritual force. Would you think of a personal God?

JB: I have problems reconciling a picture of a God who is all-powerful with a God who is loving and caring. This may be because I am thinking too linearly. If you ask me to describe an electron I will talk both about a wave model and a particle model, and can cope with that duality. But I see the loving, caring God and the God who is in charge of the world as separate, exclusive, in contradiction. It may be that my understanding hasn't moved far enough to cope with a duality in religion like there is in understanding the electron.

PC: Does your training in physics have something to do with your reticence to attribute physical events to the direct hand of an all-powerful God?

JB: That's probably true, yes. When asked about miracles, I will quite happily talk about revolutions in people's thinking and attitudes. But I find the mechanical interference of God unpalatable. I'm very suspicious of being over-ready to put God into the unknown bits or unknowable bits of a hard science. But I'm always quite happy to say, "I don't know."

PC: Do you think your ability to say "I don't know" is something that you've learned through the practice of empirical science?

JB: I'd always thought it came with maturity. If you're not desperately anxious to prove God did this or didn't do that, and can be more relaxed, then you're prepared to say, "I don't know." But if you're feeling defensive then you try too hard.

PC: So for you, part of growth as a religious person is to be silent at the areas where we encounter mystery. I wonder whether there's some aspect of your scientific training that contributes to this willingness to accept silence?

JB: Well, being an astronomer, one is dealing with a lot of things that are beyond our grasp. I would certainly say that one has to stand silent before things that you don't know in astronomy as much as in religious areas.

PC: So the success of physics doesn't give grounds for thinking that ultimately all will be explainable in physical terms?

JB: I would, on principle, be suspicious of an absolute statement like that. Increasingly I have come to understand that scientists have done a lot of damage by being overconfident. I think scientists, particularly physicists, have got to learn again that what they do is often highly abstracted, a long way from everyday life, and that what they do is often based on some fairly speculative assumptions, a fact which they commonly forget.

PC: Has your sense of God grown through the development of the physical sciences?

JB: I'm not sure that it's grown in quantity, but I think it's grown stronger. I think that some of the science/religion discussions have actually strengthened religion, and probably also strengthened science. They've made people think just a little bit harder about what they were saying and what they meant and how sure they were.

PC: Immanuel Kant is famous for having drawn a sharp distinction between the empirical world which is guided by the laws of natural science, and the realm of ethics where religious language, the talk of divinity, is crucial. He at one point called these the two kingdoms. I wonder if that's the distinction that you're implicitly drawing.

JB: Yes, something rather like that, with the addition that I've become increasingly aware that I have these two kingdoms and I'm beginning to wonder if it is right to have the separation.

PC: The strength of the two kingdoms approach is that it allows you to do science in a rigorous and careful way and, at the same time, to preserve the ethical and religious life, so you can pour yourself into the two kingdoms without reservation or conflict.

JB: Yes, quite vigorously, and rapidly become a schizophrenic [laughing]. When I was younger I was very much bothered by the question, did God create the universe? I'd be tearing myself apart thinking about this. Then I'd go into a Quaker meeting for worship and, within about ten minutes, I would suddenly realize that that was not the most important question; I was totally taken by the sense of the presence of God. All these *academic* questions just floated miles away. I think it's possibly because I've got a very strong, almost mystical sense of reality. Maybe that helps me hold the two in tension without getting destroyed by the tension.

Jocelyn Bell Burnell

Science, spirituality and religion

An exploration of bridges and gaps

Introduction

One of the basic tenets of Quakerism is that, regardless of what respected authorities say, we are enjoined to speak of our own experience of the living God, and from our own experience of the living God. Quakerism appeals to many scientists, for the openness and search that this experiential attitude implies is similar to the experimental approach of the research scientist. There are strong parallels, and in particular both require integrity. Scientists are trained to check their data, but having checked it, to respect its integrity, respect what it is telling them, neither over-interpreting nor falsifying the outcomes. Similarly in our religious life, Quakers have procedures for testing one another's religious leadings, but having checked them we are encouraged to respect them and what they are telling us.

Is there a bridge?

A bridge is only necessary when there is something to be spanned, so first I would like to explore what it is that needs bridging.

There is perceived to be a science and religion "problem." Along with other denominations Quakers will point out those of their members who are believers and scientists, and the assiduousness with which this is done belies an anxiety. What the problem is, or is supposed to be, escapes me. Perhaps it is less obvious in Britain than it is in the United States. There undoubtedly are modern cosmological issues that theology and religion still have to take on board, such as that there will be an end to the habitable Earth, that there will be an end to the habitable universe, and the re-working necessary of some of our theology if there is other intelligent life in the universe. Is this what needs spanning by our bridge? Whilst these are genuine and important issues, are they generally enough known to be the source of the problem?

There is also the view, perhaps articulated most strongly by part of the feminist movement, that sees science as a threat, and maybe also sees it as an irresponsible power and plunderer of the planet. Is this the gap that needs bridging? There is *a* gap here that needs bridging. In some cases at least this viewpoint seems to be a reaction, a fear of a (powerful) subject that some are not equipped to follow. I am always particularly touched when at the end of one of my popular science lectures somebody (frequently a woman) comes up to me and remarks that they did not expect to understand, but to their surprise they did! Both their amazement, and their sense of empowerment, are a joy to see. Unfortunately such gaps can be widened when scientists appear too confident, too arrogant, too abstract, or too detached.

The island of science

Is science in our culture an offshore island? Not far offshore, to be sure, but with clear water all round it, separating it from the shore, the mainland, which is the rest of our culture? Is it the bridge to the Island of Science that we are talking about?

I picture this island as being high at one end, sloping down to a gentle shore at the other end – a wedge-shaped profile. The high end of the island has sheer cliffs dropping into the water; the cliffs are called physics! The highest point of the island, also at this end, is crowned with a lone pine, called Grand Unification Theory. Part way along the cliffs is a cove with a beach, which provides some access. The cove is called astronomy. Most of the landing places are at the other end, where the island slopes gently into the water, although careful navigation is needed amongst some rocks in the water. This part of the island is known as earth sciences and biological sciences, and between here and physics lies chemistry. (British readers familiar with Arthur Ransome's *Swallows and Amazons* may recognize the island – a case of my cultural context showing through!)

Astronomy and morality

Are there moral issues that exercise astronomers? Yes, there are, but largely only the issues that exercise many other people doing many other jobs. Furthermore, in other occupations there can be additional moral issues. For example, archaeology was one career that I considered; if I had followed that profession, disturbing ancient sacred sites and ancient burials would have raised moral issues for me. Astronomy seems relatively moral-free, perhaps because it is relatively detached. What about unexpected spin-offs and social responsibility? True, if those pulsar signals had actually been signals from another civilization out in deep space, then there would have been important and sensitive issues to be considered. We are learning (slowly) to take care of space and other bodies in space just as we have learned the necessity of taking care of this planet. The other spin-off that I am aware of is the availability of people with high-level computing and technical skills, and this seems to me desirable.

There may be funding issues – how much money is it right to spend on such a subject? However, in Britain the money spent on astronomy is much less than the revenue raised from tobacco taxes, much less than the revenue from taxes on alcohol, and microscopic compared with the money spent on weapons of war, all of which raise more acute issues than does astronomy. In the past there have been telescopes sited in countries run by unacceptable regimes. As in any other group of people, this made some astronomers uncomfortable but did not trouble others. I believe now that all our British telescopes are in countries that have more democratic regimes. So while astronomy raises theological issues, as identified above – and there is a moral imperative to address these – astronomy does not appear directly to raise significant moral or social issues.

The interaction of science and religion

Our culture affects the way we interpret experience and the questions we ask, and also sets the context of the questions. So, in a scientific investigation, the questions that are raised may be culturally dependent. Since our gender gives us cultural baggage, it is quite likely that as well as working in different ways, women may ask different questions in scientific research.

Does our religion affect the way we do science? Margaret Wertheim in her recent book *Pythagoras' Trousers* made me appreciate that there could be a link between monotheism and the search for a single all-embracing theory such as Grand Unification Theory – that monotheism and mono-theory may be linked. If this is the case, it presumably occurs at the level of society rather than at the level of the individual, and because it is part of the cultural context presumably would affect non-believers as well as believers.

Note in passing that many physicists "believe" that the equations that correctly describe nature will turn out, when found, to be simple, elegant and beautiful. Indeed I have heard physicists say of an equation "It's so beautiful it must be right!"

So today does an individual's religion affect the way they do science? Probably not, or not in a major way. However, there may be minor influences. One of the neglected, underplayed, under-discussed dimensions of science is the role of intuition and imagination. This is the creative aspect of science, the synthesis rather than analysis, which is very important when developing models or hypotheses. We can expect this to be heavily culturally dependent, and so in the models devised we might expect to see gender differences and religious influences.

One Quaker's science and theology

Perhaps it is fair to say that British Quakers, with their emphasis on continuing revelation, are less bound to a holy book, less bound to a leader, and less bound to their historical traditions than are some other faiths or denominations. This may be one of the reasons why I feel able to adjust my theology to be compatible with my science, for my science drives my theology quite significantly.

Suffering

Once upon a time a child suffering from an incurable disease was taken into hospital. The child was no fool, and knew that one of his organs had failed. The hospital staff did a good job teaching him how to manage his disease and his medication, and gave that 10-year-old full responsibility for his life. Someone sent him a get-well card, which hoped he would soon be fit and well and "good as new," but the child knew he would never again be good as new. That child was my child, my only child. Why do we suffer? More acutely, why do those we love suffer? And why do we find it so difficult to face? There are a number of "explanations" for suffering which the churches traditionally deploy; I find none of them satisfactory.[1]

We suffer because we have been bad, is the essence of one explanation. This seems an immoral explanation as it creates guilt, feeds guilt, and invites sufferers to blame themselves in order to save God's reputation. A variant on this explanation is that even if the suffering comes patchily and unpredictably, over a lifetime we get the suffering we deserve. But this does not explain why small children suffer and die, nor why obviously worthy and innocent people suffer. It implies a vindictive God.

Another explanation is that God is in charge and has God's own inscrutable reasons for imposing suffering. Comprehension is beyond humans, and indeed only those lacking in faith would wish to know the reasons. This is actually a non-explanation, and protects God and God's reputation by keeping God at a safe distance, insulated behind a barrier.

A third explanation is the "tragedy as test" model: suffering is sent by God to ennoble. If this is the case then God gets it wrong quite often; there are people who are broken by their suffering. Note also that for God the end appears to justify the means in all three of these explanations.

My label for the fourth explanation comes from the war-time experience of food rationing in Britain when mothers would encourage their children to eat unattractive food by promising them jam (jelly) tomorrow! The "jam tomorrow" explanation for suffering promises a world to come where we will get compensation for the pains of this world. This picture implies an all-powerful God giving rewards for loyalty, rather than a loving, caring God.

I find none of these arguments convincing, and I know of no others used in the Christian church. The existence of suffering poses a major problem. The scientist, when faced with this sort of conundrum, looks at the initial assumptions, checking their validity and application. If we apply this type of analysis to the question of suffering we identify two initial assumptions: that there is a loving, caring God; and that God is in charge of the world. If we insist on holding both of these assumptions then the problem of suffering identified above seems inevitable.

There is no way of checking on these assumptions, except by requiring consistency with other areas of our experience, but my guess is that we have to relinquish the assumption that God is in charge of the world. This is consistent with my understanding of creation and evolution. It also spares me the personal anguish of saying that God is not loving or caring. I still believe in an omnipotent, omniscient God, but also believe that God chooses not to use this power, chooses not to intervene directly in this world. So God will not divert that hurricane, prevent that road accident or cause the sun to shine for the family wedding. We are God's agents in this world, and it is through our grace and creativity, our generosity and compassion, that God is active in the world. When there is suffering God suffers too.

Creation

What do we mean when we talk of God as Creator? What should we mean? Deep in worship I will use the word, not because I literally mean Creator, but because there is no other word in my vocabulary to express my relationship to God. God may well be in the creativity exhibited by inspired people. However, I very much doubt if God created the material universe; I suspect it largely created itself, following the laws of nature.

Where did the laws of nature come from, and who fixed the values of the physical constants that go with those laws? There are echoes of the Book of Job here, and just as the author of that book implied, I have to say I don't know, because I wasn't there! Some scientists have noted that several of the physical constants have to be set within very tight limits for life to be possible, and estimate that there is only one chance in a million million of that happening. In spite of these long odds it clearly has happened as life exists. So perhaps that is evidence for a designer. Personally I do not accept that argument, partly because it feels like an after-the-event argument (as do all arguments from design), but more strongly because we can only observe a tiny fraction of our universe and have no knowledge of the rest where conditions may be totally different. In other words, we are here because this is a patch of the universe suitable to life.

People keen to give God a role in creation will note that there is a tiny fraction of a second immediately after the Big Bang which physicists cannot address, and place God there. I am reluctant to put God into that slot because it feels too like another case of the discredited God-of-the-gaps theology.

Being able to admit that one does not know something, and being prepared to live on with that unknowingness, is important. Living with unresolved questions is part of life; forcing an answer to a question, for the sake of an answer, on the basis of inadequate evidence feels dishonest or lacking in integrity. If there was an initial act of creation – sometimes described as God lighting the blue touch paper and standing back – then there is no way a scientist, acting as a scientist, can find out about it. Given my suspicion that thereafter the universe looked after itself, I'm not inclined to identify God with such an event, but the honest position is to admit that it is a mystery, and likely will remain so.

So, perhaps typically for a Quaker, my science determines my theology to a significant degree. I will return in the concluding section on spirituality to consider the consequences of these (non)beliefs. Does my theology drive my science in any way, other than forming part of the cultural context as already noted?

If I were not a Quaker, or not religious, how different would my behavior be? To Quakerism I owe personal empowerment. There is a long tradition of Quaker women occupying key roles, of being accustomed to playing their part in church affairs, and of being heard. Margaret Hope Bacon in her recent book *Mothers of Feminism* shows how a disproportionate number of Quaker women in the United States held leading positions in the anti-slavery movement, the campaign for votes for women, and the pacifist movement. Our tradition enables me to push at glass ceilings, to serve as a role model and mentor, as a spokeswoman and interpreter and representative of women in science in the United Kingdom. But would I do different science if I weren't a Quaker, weren't religious? Would I work differently with other people? Would I have a different set of ethics? Perhaps, but not necessarily.

Spiritual quest

It is important to me that my science and theology fit together rationally (give or take a bit, because we are dealing with huge imponderables) and I hope the previous sections have shown how one Quaker scientist's current thinking endeavors to dovetail them. However, science-and-theology issues form only a small part of my life, and only a small part of my religious life. I can experiment with different ideas, different ways of meshing science and theology and remain remarkably unaffected, undismayed by huge changes in my world picture. Whilst attributes like academic argument, careful experimentation, clear thinking, physical insight and logic are important, I also honor and value qualities such as beauty, compassion, connectedness, empathy, genius, glory, insight, inspiration, the numinous, wisdom, and wonder.

The core of my religious life lies in my spiritual life, which adds another dimension, and moves science-and-theology issues away from the center. Confidence gained in the life of the spirit has allowed me to try out different ways of making science and theology compatible without feeling that the foundations of my life are threatened. This spiritual life has developed through attendance at Quaker meeting for worship and in the quiet, faithful waiting. There I have discovered the almost palpable sense of the presence of God, when all those present become collectively still and deeply gathered. At these times my worship goes beyond words, and just being present, just breathing become my prayer and my worship. Whilst being acutely alive it also comes naturally to be hushed and full of wordless praise. One may feel known and judged, but also affirmed and assured. There is a timelessness about it. In particular, I have learned that questions that an hour ago seemed very important (e.g. in what sense is God the Creator?), float away as unimportant and irrelevant.

Conclusion

I am fully aware that what I have said may come across as irreverent, brash, immature or even mad. I plead the best of intentions and a searching struggle, a search that still continues. Perhaps only a Quaker would be audacious enough to attempt something so ambitious. Our attitude to authority, our emphasis on continuing revelation, combined with our awareness of the Holy Spirit as well as the Word give us a unique perspective.

In 1986 I was one of a small delegation from the British churches that visited the churches in the (then) Soviet Union and met with the Soviet Government Minister for Religious Affairs. At one point in that meeting he said to us "We know that God does not exist. We have sent satellites up into space and He is not there." One smiles at that, for the image of a white-bearded God sitting on a cloud went out some time ago! Will future generations smile at our world picture? I suspect so, for it seems a while since it has changed, and it seems some evolution is due.

Note

1   For a more extended discussion than I offer here, see my book *Broken for Life*.

References

Bacon, Margaret Hope (1997) *Mothers of Feminism*, Philadelphia: Friends General Conference.

Burnell, S. Jocelyn (1989) *Broken for Life*, London: Quaker Home Service.

Ransome, Arthur (1930) *Swallows and Amazons*, London: J. Cape & H. Smith.

Wertheim, Margaret (1995) *Pythagoras' Trousers*, London: Fourth Estate.

2   Kenneth S. Kendler

Kenneth S. Kendler is the Rachel Brown Banks Distinguished Professor of Psychiatry, Professor of Human Genetics at the Medical College of Virginia, and Co-director of the Virginia Institute for Psychiatric and Behavioral Genetics at Virginia Commonwealth University. He has published more than 300 articles and several book chapters and has received the Lieber Prize for outstanding research in schizophrenia as well as the Dean Award for 1998 from the American College of Psychiatrists in recognition of his major contributions to the understanding of schizophrenic disorders.

Interview by Gordy Slack

GS: Would you say a few words tracing the history of your own religious practice?

KK: My upbringing was culturally very Jewish but anti-religious. My parents come from that division of Ashkenazi or European Jewry who'd been socialist for three generations. They were political in orientation and both anti-religious and anti-Zionist. In late adolescence I read a fair amount about Eastern religion, particularly about Zen Buddhism and not a great deal about Western religions, which I think was not atypical growing up in Santa Barbara in the mid-1960s. In my second year of college, when I was trying to obtain some focus about my academic and personal goals, I was offered the possibility of going to Israel on a six-month field study.

It was a seminal experience. I spent the first three months studying on a kibbutz and had a wonderful time. I got to know and came to deeply admire a number of the kibbutzniks there. I learned Hebrew relatively well, and began studying the Bible. During the last three months I decided to get more specific religious training. I was not seriously considering converting to being an Orthodox Jew, but I wanted to learn about that world. I had a quite intense experience for three months, living the life of the Yeshiva student.

After I got back, I had the task of integrating that experience. It hasn't been terribly easy. Orthodox Judaism was never very viable for me, in part because of the strong upbringing I've had in Western scientific and intellectual pursuits. In the Orthodox framework you need to accept that the original Torah was given by God to Moses on Mount Sinai and that all of the law that derives from that has the same divine commandment. Although I can respect that, it was clear to me that I couldn't personally accept it.

Reform Judaism, now my formal affiliation, is only marginally satisfactory to me. In many ways it's watered-down orthodoxy. I have set myself on a trajectory which is largely personal and scholarly in its orientation. Since leaving college I have made a serious effort to study the Hebrew Bible. I kept up my Hebrew and I have been slowly working my way through the books of Genesis and more recently, First and Second Samuel. I've also been doing some teaching at our synagogue. And I've been doing a fair amount of study in ancient near Eastern religions. I've recently been studying Mesopotamian mythology, particularly creation stories.

My other religious strain is the Eastern religion connection. In college, before I went to Israel, I spent a summer studying at the Zen Center in San Francisco, a very high quality institution. I meditate every day I can, in the morning. It is partly an attempt to focus myself on the fundamental aspects of each day, of trying to appreciate each day as it comes, trying to collect myself and prepare myself, but there is also a religious component. When I got back from Israel I tried praying in a more traditional Jewish way but for me the silence of zazen is a more profound form of prayer than a traditional Jewish Orthodox liturgy.

GS: Do you find that your religious study and practice motivate the direction or the methodology of your scientific research?

KK: For the most part I have existed on a two-track mental approach. I do research and I pursue my religious thought and practice largely independently of one another. I'm not sure the solution is entirely satisfactory, but it does have a conceptual basis and it does stem partly from being a psychiatrist. You can use a variety of different words for it. One set of words is "knowledge" versus "wisdom." The science that I pursue is on quite a different epistemological level than the religion I pursue, and the two don't use similar methods, don't have similar goals, and in some substantial way don't conform to one another. Maybe the best way to put it is to say that they complement one another. They don't conflict, but they don't entirely exist on the same plane.

This does relate to my attempt to make sense out of my identity as a psychiatrist, because part of what you do in clinical psychiatry is try to integrate the scientific knowledge we have about psycho-pharmacology and about neurotransmitters and brain function with the very human knowledge that you gain about people as you sit with them in a counseling or psychotherapeutic endeavor.

It's not an accident that I've taken this route rather than become a biochemist or a cancer geneticist, and it does represent my struggle to compromise my scientific goals with my religious goals. It's been a bit of a devil's bargain for me; I write papers about critical human issues like the nature of the parent–child relationship, about how we cope with adversity, about psychosis, which is the fundamental ability to perceive accurately social reality. The nature of my science is fundamental human issues, which undoubtedly are of interest to me and related to my religious urge to understand the nature of the human being and the human condition.

Intriguingly, when I was a medical student at Stanford, the very first study I ever tried to do in psychiatry was to look at people that had very severe forms of non-Hodgkin's lymphoma, who had on the average about four months to live. I wanted to examine the relationship between their religious beliefs and their coping strategies in their last months of life. But the oncologists refused to let me talk to their patients. The situation hasn't improved much since then. Given the importance of religiosity in human behavior, it's appallingly poorly researched in the field of mental health.

GS: Psychiatry and religion have been seen by so many as competitors for a long time.

KK: There have been some hostile relationships. Freudian influence is fundamentally anti-religious. There is still literature suggesting that religiosity is a bad thing for mental health. This is not what we've found. We found, for example, that internal religiosity – not so much the external form of being ritually compliant or ascribing to more fundamentalist Christian beliefs, but the internal sense of feeling a religious purpose or focus in your life and praying to God at times of stress – was a fairly good protector against the depressogenic effects of stressful life circumstances. That's very consistent with what I would have predicted overall.

GS: Going back for a second to the distinction you made between knowledge and wisdom, do you see those two objects of your study as having a singular source?

KK: Oh no. Part of what you learn by studying the history of psychiatry is the terrible things that well-intentioned people can do to other people in the name of what they think is medicine. That's particularly true in psychiatry because we always attribute all the good things to things we've done, and all the bad things to the accidents of nature or the uncooperativeness of the patient. In consequence, I try to be very hard-nosed about the scientific method. The only way we are going to advance this field is by applying the most rigorous kinds of conceptual perspectives. There are people working solely from the perspective of family dynamics who think that everything is the result of geneagrams and cultural transmissions from grandparents. Then there are the social psychiatrists who think that everything is a result of poverty and oppression. The biological psychiatrists say everything is due to neurotransmitter abnormalities. All of them are ideologically driven.

My identity within the field of psychiatry has been tied to saying: "Look guys, we have to treat human behavior the way we treat other things: scientifically. You may see it as bloodless – and I'm not saying I care for patients this way, because I don't – but you need to state your hypotheses and be clear about the statistical methods you use to address the hypotheses." That's what I mean by "knowledge." Knowledge is something that is ultimately testable, although there are times in the field of psychiatry when it's damn hard to test and to replicate and still get what you mean across.

I think that human wisdom comes in many, many different varieties. Human culture is very diverse, and in fact arrives at fundamentally different views, though there are some common connections. Part of wisdom means facing the rather negative features of the human character; we're pretty aggressive. This is something that as a Jew you have to face. The human capacity to see somebody who's not a member of your group as not human is frighteningly profound, and has had tragic implications throughout our history. Certainly there are aspects of the wisdom of Judaism that I find very appealing, but some of it is very close-minded, very ethnocentric. But there's a tremendous amount of beauty in the tradition.

My Zen path is also active. I read Tang Dynasty poetry, haiku and some Zen philosophy, but mostly what I do is meditate. I certainly feel that there's not one way or one path. I think part of early phases of religious development in most cultures is to feel that you are the chosen people and your capital is the center of the universe.

My religiosity has been a private journey, which is clearly connected with the latter part of the twentieth century. This is a hard time to live, in some ways. We're moving so fast. Families are torn asunder. I'm scared sometimes of losing hold of what I'm doing in this life. What Judaism represents for me is a powerful sense of continuity. When I sit down and study these texts which not only my grandparents, but their grandparents, and their grandparents before them, back twenty-five hundred years have studied and tried to make sense of there is a powerful sense of belonging to something that is orienting. I'm frightened for my children – Nintendo, the computer, TV – how will they learn to center themselves? I don't think we were meant to live this way. We were meant to grow up in smaller communities with a much greater sense of continuity and structure.

Kenneth S. Kendler

Must not the judge of all the earth do justice?

God's nature and the existence of genetic diseases in man

Introduction

The goal of this essay is to review my struggle to make sense of two contradictory propositions, each of which I believe.

1   God is fundamentally loving and cares deeply about the welfare of humankind.
2   Genetic mutations arise through random chemical processes and are then, through the random process of genetic segregation, transmitted across generations and predispose to psychiatric and substance use disorders which in turn cause terrible suffering to the afflicted.

I divide this essay into five sections. In the first, I outline my background and the attitudes toward religion and science with which I began this process. In the second section, I try to describe the nature of the central problem of this essay as I have experienced it in my professional and personal life. The third section outlines three biblical "responses" to this problem. The fourth section outlines several other possible responses of which I have some knowledge. The fifth and final section presents a summary of my current thoughts about this issue. Let the reader be warned that I will present no fully satisfying solution to this wrenching problem.

My background

I am trained as a physician, psychiatrist and geneticist. The focus of my professional career has been the study of the genetics of psychiatric and substance-use disorders. However, I am also an active clinical psychiatrist and therefore care for patients suffering from the same disorders that are the subject of my research. While this is not the place to review the evidence in detail, we now know, with some considerable confidence, that genetic factors play a substantial etiologic role in all major psychiatric disorder such as schizophrenia, bipolar affective disorder, anxiety disorders, alcoholism, and drug abuse.

I was raised in a family that, while culturally very Jewish, was non-observant. I had no formal Jewish education during my childhood or adolescence. For at least three generations, my Jewish ancestors have been more concerned with social justice, the labor movement, and assimilation into American society, than with the transmission of the Jewish religion.

As part of a process of self-discovery, at the age of twenty, I spent six months in Israel. The last three months of this visit was spent living the life of an Orthodox Jew, studying in a beginner's Yeshiva, an institute for traditional Jewish learning. From that time began a journey of trying, across a gap of three generations, to reintegrate Judaism back into my life and the life of my family. My family belongs to a Reform Jewish Synagogue, celebrating the Sabbath and the major Jewish festivals. I am not Orthodox and therefore do not see the corpus of Rabbinic law, the Halacha, as binding upon me. Probably of the greatest relevance for this essay is that I made, since my college days, a life-long effort, as an amateur, to study the Hebrew Bible. This course of study, which I pursue with regularity on Friday evening, is both a scholarly and a personal journey.

When it comes to knowledge about our world and how it works, I have always believed that science is our best and least fallible source of knowledge. I would have sided with Galileo and later with Darwin in their struggles with religious authorities. On the other hand, science never seemed to me to be a source from which to obtain wisdom, guidance on ethical questions, or answers to ultimate questions about the nature and meaning of human life. Evolution might explain our emergence from the anthropoid apes in a mechanistic way. But, I would also believe, as the author of Genesis says, that we humans are special. On the one hand, I can explain this specialness as an emergent property of an increasing cerebral cortex driven by selection for language and symbolic reasoning. On the other hand, I can see this as the author of Genesis (1:27) does:

> And Elohim created man in his own image/likeness, in the image/likeness of Elohim, He created him.
>
> [all translations mine]

While our emergence as a species is explicable entirely on the basis of well-understood evolutionary principles, humans are also touched by God in a way that sets us apart from other species of the world.

The problem

Suffering

I have, in my medical training, watched people suffer and die from a range of medical diseases: heart disease, diabetes, cancer, stroke. They are all terrible in their own way. Few, however, compare with the grief caused to the sufferer and his loved ones by severe psychiatric illness. Let me briefly review a typical case of schizophrenia (with details changed to protect confidentiality), amongst the many for whom I have cared.

John was seventeen when I first saw him. Up to the age of fourteen, he was an out-going, even-tempered, bright young man. He had done well in school, had a range of friends and liked sports. He was the kind of child that parents would wish for. However, a few months after his fifteenth birthday, John began to change. His ability to concentrate in class diminished. His interest in his friends declined and his increasingly odd behavior alienated those who had previously been close to him. His personal hygiene began to deteriorate. As his grades declined, more and more notes were sent home with him from school. A series of teacher–parent meetings were held. His parents were told he needed to apply himself better. Tension increased between him and his parents as fights occurred over whether homework had or had not been done, whether he had taken a shower or not, and when he would do his household chores. His parents were reassured – "This is just a phase," they were told. However, his behavior became more and more peculiar. He refused altogether to bathe. Apart from school, he spent all his hours in his room listening to rock music. His parents had a more and more difficult time understanding him and tolerating his behavior. A few months after his seventeenth birthday, John came down from his room to announce to his parents that he would be leaving home the next day and joining the world tour of a famous rock and roll band. He informed them that he had been receiving secret communications from the manager of this band from the back of his telephone for many months. In recent weeks, the manager and several members of the band had been talking to him regularly, at first just at night-time, but more recently most of the day.

When I saw John, he looked nothing like the bright-eyed adolescent I could see in the family pictures the parents brought along. He was mumbling to himself, obviously actively hallucinating. His mood was flat and he was distant as if living entirely in a world of his own with little interest or time for the real world around him. He spent most of the day staring at the wall or floor mumbling to himself, preoccupied by a world of grandiose fantasies about being a famous rock and roll musician.

The agony in the faces of his parents when I spoke to them about the nature of the schizophrenic illness of their son is hard to describe although I suspect those who have children of their own can imagine. Yes, the mother said, she had an aunt who had spent all her adult life in an institution. She remembers the name schizophrenia being applied to her. "But my husband and I are well. Our three older children have all been fine. Why did this happen to John? He was such a lovely child." The mother began to weep openly. John did not respond well to our current available treatments. When I lost touch with him two years later, he was a patient in the chronic wards of a state mental institution.

Part of my research involves studying families like John's which contain two or more individuals with schizophrenia. In all these families, I have listened to well members talk about what can only be described as survivor guilt.

> I think, doctor, about my own life, the fact that I have been able to marry, to have children and a good life. And my poor brother, he has none of these things. What did I do to deserve this happiness and why has my brother had to suffer so much?

The nature of genetic transmission

The basic features of genetic transmission are too well known to be worth reviewing here. The key process is genetic "segregation" during meiosis, the process wherein the developing egg or sperm is given one of the two available copies of each chromosomal region. Aside from a few minor exceptions that need not concern us here, this process is as close to truly random as anything we know of in the world of biology.

Part of my research work has been to localize, at particular places on the human genome, the specific genes that predispose to schizophrenia. In my role as a scientist, I accept, and indeed deeply believe in, the Darwinian, molecular-genetic paradigm for the evolution of all life on earth, including humans. This means that I also believe that, at some point during the evolution of our species, genetic variation emerged via random mutational events that rendered the carriers of those mutations at greater risk for schizophrenia. For reasons that still remain unclear, that mutation or group of mutations has remained at reasonable frequencies in our population. And these mutations pass themselves randomly through families like those of John.

The loving God

From my first experiences in my adolescence of what I felt to be the divine, a central feature of that force has been its goodness and care. This has been relatively easy to synthesize with my growing knowledge of Judaism. Our history begins with a personal, caring relationship between the patriarchs and Yahweh-God. This relationship is then expanded to include God and all of the people of Israel. Finally, in parts of the prophetic literature, this is expanded to include God and all of humankind.

God, in the Jewish tradition, is intimately involved in the lives and actions of humankind. In the traditional "eighteen benedictions" that constitute the core of orthodox prayer, we are to take three steps back to remove us from the human world and then three steps forward to come into the direct presence of God.

We praise God, but we also ask things of Him. For example, one of the prayers on Rosh Hashanah, the Jewish new year, reads:

> Our father, our king, listen to our voice.
> Our father, our king, keep far away from us pestilence, the edge of the sword, and famine.
> Our father, our king, write us in the book for a good life.

In Judaism, God is not a distant judge. The relationship between God and humanity is an intimate one. Indeed, a long Jewish tradition teaches that humans and God co-operate in completing and fulfilling the promise of the creation that God began. This partnership is clearly seen in the second chapter of Genesis (2:19):

> And Yahweh-God made, from the earth, all the living things of the field and all the birds of the heavens and he brought them to Adam to see what he would name them. And all that Adam called them, each living soul, that was its name.

The relationship of humanity and God is an intimate one – partners together, naming God's creatures and watching over them.

One of my favorite of the many blessings contained in Judaism, initially said by the priests over the Jews gathered in the Temple grounds, but now said by a father over his children on Friday night, is this:

> May God bless you and keep watch over you. May God make His face to shine upon you and be gracious to you. May God lift up His face before you and may He give you peace.

Nothing that I know brings a human further from this peace, the greatest of blessings, than severe mental illness.

The just God

God within the Jewish tradition is deeply concerned with justice. Biblical passages far too numerous to mention deal with this theme. Most typically, it is expressed in the injunction for humans to act justly, as we see, for example, in the teachings of Amos (5:14–15):

> Seek after good and not evil in order that you may live … and establish justice within your gates [the place in the cities of ancient Israel where all legal transactions were carried out].

The central problem addressed by this essay is my inability to understand how a God who is caring and loving, who is a partner with us in creating and sustaining this beautiful and wondrous world, who values justice so highly, could inflict upon individuals, in an apparently arbitrary fashion, these terrible psychiatric disorders.

Three biblical responses

I now want to review three responses to this problem that come from the Hebrew Bible. I will describe them in what might be reverse chronological order: Koholet, Job and Avraham.

Koholet

In the ninth chapter (v. 11) of Koholet (Ecclesiastes), we have the following famous quote:

> I turned and saw that under the sun not to the swift is the race, and not to the strong is the battle, and also not to the wise [is given] bread, and not to those of understanding [is given] riches, and also not to those of knowledge [is given] favor; for time and chance [or happening or occurrence] befalls all.

Koholet's solution is that indeed, chance governs the universe. It is naive to expect virtue to be rewarded or justice to be served. Genes that predispose to mental illness will be passed on to some individuals in a family and not others because that is the nature of the world. If there is a loving God or a God that values justice, He must be acting at some other level.

Job

After questioning God for many chapters, trying to understand why, despite his innocence, Job has suffered so much, God finally, in his awesome speech from the whirlwind (chapters 38–39), gives Job an answer. This speech is so famous and long that I will only paraphrase God's response here.

God gives Job a tour of the wonders of His created universe, repeatedly asking him where he was when the world was formed, whether he can perform the many duties of God in the creation and sustenance of the world. The net effect of this overwhelming vision is to communicate to Job that God is beyond our power to understand or judge.

Job is a profound book that can merit a lifetime of study. While I have far from a complete knowledge of this work, it seems to me that the author is trying to tell us that a complete understanding of God's universe is simply beyond our capacity. Therefore, we cannot judge God's actions and cannot apply to them human standards of right and wrong. According to the author of Job, I cannot conclude that the suffering of those afflicted with schizophrenia is arbitrary, without meaning or justice, because there are mysteries in God's universe that are far beyond my comprehension.

Avraham

While God often demands human justice in the Hebrew Bible, so – in their partnership – can humans demand justice of God. This is no place better illustrated than the dialogue between Avraham and God about the fate of Sodom in Genesis 18. Avraham asks the question of God: "Will you sweep away the righteous with the wicked?" Avraham goes on to tell God the answer that he expects (18:25):

> Far be it from you to do such a thing – To kill the righteous along with the wicked! That the righteous should be [treated] as the evil – far be it from you! Must not the Judge of all the earth do justice?

Yahweh-God evidently agrees with Avraham and states that He will not destroy Sodom if it contains fifty righteous individuals. Avraham and God then begin their famous bargaining session. Avraham succeeds in getting God to agree not to destroy Sodom if only ten righteous are found within its walls.

The message here seems quite clear and very different from that conveyed in the book of Job. God's moral actions can be judged by human standards. God cannot be allowed to mete out punishment without justice and humans have a place in calling God to account for His actions.

Other possible solutions

I am trained in neither theology nor philosophy, so my review of other extra-biblical solutions to this central dilemma will be incomplete and superficial. However, I need to briefly review four.

All is made good in heaven.   A solution to innocent suffering, more prominent in Christian than Jewish thought, is to postpone God's accountability to the afterlife. Innocent individuals who have suffered with psychiatric illness will be especially rewarded in heaven.

Quantum mechanics.   We know that, at the level of subatomic particles, indeterminacy and "chance" hold sway. Since original mutations are the result of errors of an enzymatic process involving hydrogen and covalent bonds, mutations can be seen as an inevitable part of a fallible chemical process. Given the current physical construction of the universe, God could not, even if He wanted to, design a system for the replication of the DNA molecule that would be without error. The fact that one individual in a family gets schizophrenia and not the others is just due to the "bounce of electrons."

Small evils for the greater good.   God could be seen as maintaining His fundamental goodness if the process of mutations in DNA and the diseases that they cause were a small evil that was part of a larger good. If evolution is seen as a great good, it follows that the processes of mutation and the subsequent natural selection against deleterious mutation carriers are part of the package. If you want to have the good of evolution, you have to suffer these lesser evils.

God the clock maker. Another approach related to the "small evils for the greater good" solution is the deistic model of God. He set up the evolutionary process, perhaps with an inherent goal of producing intelligent life, but perhaps not. Then, He lets the process unfold, playing no role in the day-to-day struggles of "nature red in tooth and claw."

Conclusion – from independence to dialogue

When I was first contacted to participate in the Science and the Spiritual Quest Project (SSQ), my religious and scientific selves existed in separate domains. I compartmentalized myself. The gross contradictions between my sense of God and the senseless tragedies that I saw before me in my clinical and research work went largely unnoticed.

For better or worse, this divide has been lowered by my involvement in this project and the preparation of this essay. Let me now try to review where I stand in this evolving process of trying to bridge the gap between science and God on the central issue of this essay.

In discussing this problem with friends and colleagues, variants of the Koholet solution were most commonly proposed. "Of course the world is just a jumble of molecules. All is chance. You are crazy to expect that any of it will make sense or be ethical. Get real!"

There are times that I am tempted by this solution. I do not "know" that God exists the way I "know" scientific facts. But there are other ways of "knowing" and my sense of the presence of divinity in the world is too strong to deny easily.

The closely related quantum-mechanics solution is no more appealing than that of Koholet. It only appears to solve the problem of innocent suffering. It gets God off the hook on a technicality: "He really would like to make a just world, but the subatomic particles would not co-operate."

I am also clear in rejecting the "All will be made good in heaven" solution. It is too easy and reflects an abdication of accountability in this life, the only one we know. It is also, at least in my view, not a very Jewish solution. We are stuck with the one world we know and have to try to make sense of it as it is.

Three of the proposed solutions, which I would see as inter-related, have some appeal for me. The book of Job, the "small evils for the greater good," and the deistic God all have the common element of a "Designer" God, who has established a natural universe of awesome power and beauty with aims that we see but darkly. Having established it, God lets it take its course, realizing that there will be winners and losers and that things will not always be fair.

The process of evolution is profoundly powerful and beautiful. Although I do not believe in teleology in biological systems (e.g., God did not set up evolution so as to produce *Homo sapiens*), I can nonetheless dimly see how a benign God might establish a universe within which evolution could operate and then stand back. And yet, God would have to be hard-hearted not to see the suffering.

However, my sympathies lie most closely with Avraham. This is to me the most Jewish of the solutions. Somehow, we have to stand witness. While God may be mysterious and we unable to understand His ways, yet we still have the right to ask, and the duty to try to understand. The relationship between humans and God is not one-sided. He cannot expect us to try to live our lives seeking justice and not be asked by us the same thing in return. It is not blasphemy to question and seek to understand, to expect and hope that the Judge of all the world will do justice.

I am painfully aware that this position is, in some important ways, no solution at all. It does not explain the random suffering of those with mental illness. And yet, it is the best that I can now do. If I were to give Avraham's position in my own words, it would be something like: God, I cannot understand why you permit these terrible things to happen to humans that You are supposed to love. Perhaps there is a wisdom in Your plan that I cannot see. But I challenge You to explain Yourself. I cannot and will not accept this as simply "part of Your plan" or "all for the ultimate good." I will not curse You if You do not respond, but I await Your answer …

Postscript

One of the consequences of my experiences with SSQ is that I have been more willing to discuss religious issues with my patients. One of them – we will call him Jeff – developed schizophrenia during graduate school. Although he has been free of major psychotic relapses for nearly five years, he is well aware of the limited life that he leads compared to his classmates at graduate school. He is supported by disability insurance and works four hours a week in a volunteer job at Red Cross. When he gets upset, he still feels overwhelmed by the thoughts of thousands of people rushing into his mind, like telephone calls to a broken-down switchboard. His previous psychotic episodes had all centered around his deep desire to establish a love relationship with a series of highly competent women that he met in college and graduate school, a goal for which he was poorly equipped psychologically.

However, four years ago, at a half-way house, he met Kathy, a mildly mentally retarded young woman with intractable epilepsy. They fell in love and Jeff now lives with her. With much tender care and patience, he takes care of her. He bathes her, brushes her hair and teeth and helps change the sheets when she urinates on herself during her frequent nocturnal seizures. I knew that he read the Bible and we spoke occasionally of our mutual interest in this book. I recently asked him if he thought about why he had developed schizophrenia. He replied,

> Yes, I have thought about that a lot. At first, I was mad at God because I know all the things that I have missed, the job that I had been trained for all those years in school, a normal wife and kids. I couldn't make sense of it. But, in these last few years, I have come to feel that I was meant to get schizophrenia. God had something to do with it. If I had never developed schizophrenia, who would have taken care of Kathy?

3   Kevin Kelly

Kevin Kelly is a co-founder and Editor-at-Large of *Wired* magazine, which won the National Magazine Award for Excellence in 1994 and 1997 during his tenure. Kelly is also member of the Global Business Network, a distributed think tank. He is the author of *The New Rules for the New Economy* (Viking, 1998) and *Out of Control: The New Biology of Machines, Economic and Social Systems* (Addison-Wesley, 1994). He is currently heading up a project to inventory all living species on Earth.

Interview by Gordy Slack

GS: Would you say a few words about your own religious background?

KK: I would describe myself as a postmodern evangelistic Christian. My core beliefs are fairly orthodox, but I express them with the perspectives of science and other current intellectual thought. It's my way of saying that within what might be considered my fairly classical orthodox view, I find great room, and great intellectual power, and a way of seeing the world today.

GS: "Postmodernism" suggests the embrace of multiple points of view, and yet "evangelistic Christian," at least on the surface, suggests a strong allegiance to one formal point of view.

KK: I don't mean the usual deconstructive theory. I use "postmodern" because "modern" is a word that doesn't mean anything [laughter]. Let's call it "Millennial" or "Wired" evangelistic. All I'm trying to indicate is that every age helps us to better understand things we knew before. Our understanding of science, I think, can amplify some of the ancient truths.

GS: Technological and scientific progress depend on figuring out new ways of looking at things, of finding new approaches to old problems. Religion, on the other hand, has a bias toward tradition. Do you feel a strain there?

KK: Not too much. Christianity, especially in my reading of it, is primarily concerned about human nature and its consequences, which I think most biologists would agree has not changed that much in the last few thousand years. The insights of Christianity are still applicable, and will be for a while. I think science is tremendously effective in understanding human nature and the natural world. There's no comparison between what we know now and what we knew before in terms of sheer depth and volume. Science and religion are looking at two facets of our lives and they inform each other. They interact with each other; they can sometimes cause tension. At other times, they can cause inspiration. But I think anyone who reads the Bible to understand natural history, to understand how the material world works, is not going to get very far.

It's also obvious that if you go to science for your theology, to help figure out what kind of a person to be, that won't get you very far either. But more importantly, from hanging around scientists (I'm also married to one. I'm a science groupie), I've learned that the way we make science is not objective. It's driven by questions that we ask. It's driven by preconceptions that we have, by hunches, by imagination. Our religious beliefs feed all of those things. Science and religion are directly connected in that way.

GS: When you think about the development of artificial life and artificial intelligence do you feel like religion has anything to say about what would be best for people in the application of these technologies?

KK: I believe there is a Creator who made something out of nothing, and the most remarkable achievement of this something made out of nothing was that, over time, it increased its somethingness to the point where there was what we call life, and that this somethingness increased to the point where we have something called mind. To me, the greatest "miracle" at that point is that the something recapitulated the act of the Creator in creating something itself. We have people walking around making things out of nothing, creating somethings where nothings were before.

We're now at a second, very interesting juncture where the computers that the humans made may get to the point where they also can imitate God and have their own thoughts. The things that we're creating are very much like life itself. We have to at once find them beautiful and inspiring and lovely, and at the same time, we have to really work to keep them in control. Like with children, we have to discipline them as well as give them freedom. I treat artificial life almost as if it were real life.

GS: The traditional religious interpretation of life implies a definite sense of purpose, both to human experience and action and to the way things are unfolding on a broader scale. But so many twentieth-century scientists insist that the universe is not the result of God's intention, and that to say that it is ignores one of the universe's essential qualities: its purposelessness. How would you frame that debate and where you fall in it?

KK: I think you can acknowledge purpose in some of these systems and not be blind to how they work. Everything that I've seen about these very large systems that we create is that we see that a purpose can emerge out of purposeless parts. A lot of people find it incomprehensible how a bunch of little parts that have no purpose whatsoever can be hooked together and produce something that has a purpose. I think the next millennium, if not the next century, is going to produce systems that do these things, and I think people will be converted to believing that this is happening very quickly once they are confronted with machines that do these things.

I think a lot of our understanding of God from now on is going to be informed by us being like gods. As we attempt to create life, as we attempt to create other minds, and maybe as we attempt to create other universes, we will have to understand how to be gods. This forces us to become theologians. It's like how the Web forces everybody to be a librarian, to think about things that no one thought about for a long time – indexers and all that other esoteric stuff. I think theology has a future because we're all becoming second-hand gods.

GS: Would you mind saying a little bit more about your own religious history: were you born into a Christian context, or did you come to religion later in life?

KK: I grew up trying to be the world's best altar boy and sort of then became a science groupie, and couldn't decide whether to go to art school or M.I.T., and then instead went to Asia as a photographer. I spent ten years roaming the most remote parts of the world in a time machine, basically stuck in the fifteenth century, in medieval ages.

GS: Would you have described yourself as a Christian during that period?

KK: No. I would not. At the time, however, I had a high school friend who had converted to Christianity. He was a fanatic. He was always quoting the Bible at me, which I couldn't stand. He would give me things to read, so on the backs of buses I was reading the Bible to be able to at least refute his outrageous statements. I found it really fascinating and read it for years as I was traveling and saw people actually worshiping a golden calf. I saw people making sacrifices. I saw all the things I was reading about. It became very real. I happened to be in Jerusalem on Easter week and for reasons beyond my comprehension, standing in front of the empty tombs on Easter morning, I believed that Jesus was God. So I stopped traveling and came back and graduated.

GS: That moment of realization was something that happened to you, more than a decision that you made?

KK: One is never clear about these things, but no, I did decide that I believed. It was a surrender to something that I had been thinking about. As a photographer, one of the most colorful aspects of culture, and particularly these countries, was religion. I was at every religious ceremony. I had been in every religious house of worship. I was fascinated. I photographed and read. I knew a lot about these religions in a deeper way, not just theoretically, but how people actually lived. I was very impressed with the sincerity and devotion of people around the world. The claims of Christianity as being the Truth were very hard to reconcile as I was traveling because everywhere I went was evidence of people's devotion. I had been wrestling with that in my mind all along. I was there emotionally much earlier than I was there intellectually. But in the end it was an intellectual conversion.

In my view, the logic of God and the logic of no-God are both presented by the world. You can look at the world and logically deduce either. So in the end it comes to adopting one or the other viewpoint. After thinking about this, there was more to be gained by adopting the viewpoint that there was a Creator and His interaction with the world than there was otherwise. It just happened that the trigger for this was very emotional, but it was an intellectual stance.

I actually think that we're coming to the phase where an understanding of God will become essential to a further understanding of the universe. I can adopt the Godless viewpoint as an intellectual exercise. But I'm usually aware that this is not my ultimate foundational stance. To me, the ultimate foundational stance is that a more complete understanding (I don't know if we'll ever have a totally complete understanding) of how the world works will incorporate knowledge of a Creator and His interaction with that world.

Kevin Kelly

Theology for nerds

Introduction

I don't really like computers. I find them frustrating and terribly unfriendly, but they *are* very handy, so I continue to use them, as do many others. In this essay I would like to present what I see as an emerging sense of theology coming from the street, from people I interact with, those who make technology – the nerds of our day – and are changing the way we would view the world.

In the old view of the cosmos, we humans were at the center. Science and faith were closely united in this view. People had very little problem holding both in their heads at the same time. But starting with Galileo and Kepler, and eventually Darwin, there was a great upheaval in the way we understood the cosmos and ourselves. There was a discontinuity, a sense in which we were displaced from the center of the universe. Darwin came along, Freud a bit later, and with them came a shift in our ability to give a unified explanation of the world. That's where we are right now. The consequences are still flowing from this shift.

However, even as we continue to deal with these consequences, something just as big is happening again. It doesn't really have to do with the turn of the millennium. It doesn't even have much to do with people's interest in religion – these things come and go throughout history. Sure, they are part of it, but something actually much more fundamental is happening. A cultural shift is overturning our views of faith as much as in the days of the scientific revolution. Ask yourself this question: What kind of event would cause the greatest questioning among people of all faiths? My answer: aliens.

If we have contact with an alien, it would be a tremendous event. Think about it, *Contact* was the first movie I've seen in a long while where the star was a theologian. If E.T. suddenly showed up, everyone would be rushing to theologians to help them explain what was happening. Of course, this is not something we can count on, not something we can create, but we do have something very much like aliens in our midst: robots. OK, they aren't much like aliens, not yet anyway. But the impact of the things we are creating, the things we are going to create, will be like encountering someone in outer space, like meeting E.T.

Regenesis

Are robots really emblematic of this cultural shift? Will the robots and artificial intelligences (AI) we are trying to build really force us to re-examine all
kinds of theological issues? I think so. Isn't it curious that for seven dollars
you can attend a great theological conference just by going to a movie put
out by Hollywood? Movies like *Contact* are morality plays, rehearsing all the
basic questions people grapple with: What are humans for? How does God
relate to humans? Where does God fit into our world? Where do we fit into
the universe? More than anywhere else, questions like these are actually
being asked in today's science fiction and special effects movies. They give us
a kind of popular theology.

I like to call it "regenesis." What we are doing with robots and artificial
intelligence is remaking ourselves. We now have the power to create, as God
did, and then to let our creation go. We are making things that think, that
can create themselves. We are making little worlds that run by themselves.
We are making artificial worlds, with artificial light, that we can climb into.
All these things taken together, driven by the power of technology, are
what I call regenesis. Right now, we are being confronted with the power to
make other worlds, like God made the universe. This power raises serious
theological issues about God, and about our own role in the universe.

I came upon this idea by talking with computer scientists and others who
were involved with AI systems. As I listened to them talk about what they
were doing, I was struck by the language they used. Those who spent time
on the Internet in multi-user domains (MUDs) referred to the MUD
masters, those who controlled these virtual worlds, as gods. To these people,
"God" was a real, finite, definite entity. These gods had certain powers and
abilities, and the people who played these games were very comfortable with
this idea. One of the MUD masters commented, "When I play God, I crank
up the game rate to increase the rate of evolution." He knows he's being
God, doing the kind of things God would do. Here is someone else who is
completely in charge: "Even if my world gets as complex as a real one, I'm
still God." He knows that his own power is real, not abstract at all. When
there's all kinds of havoc going on, you want to be a god, not a person.

As a result of these conversations, I became interested in the different
ways the word "God" was being used. Clearly, not everyone meant the same
thing. It occurred to me that thinking about the different ways in which one
can be a god in these games might be helpful for thinking about the one God.

The third culture

With all that computers and other kinds of technology have enabled us to
do, we are witnessing the emergence of a new culture of "spirit." This
culture is different from that of either science or art. It is the culture with

which I am aligned at *Wired* magazine, and it has some characteristics that are important to understand.

An artist, maybe even a theologian, thinks about ways to express the human condition, ways to investigate who we are. This involves lots of interior work, lots of getting down to the essentials. In other words, you might choose to think about the mind by exploring how you think; you might tell stories about how people use their minds, or about how people relate to each other in their minds. Scientists explore the mind by doing controlled experiments. They try this, prove that. The interesting thing is that a nerd explores a mind by trying to create a mind. This is the model of the third culture. You make things. The answer to your question is a technology or a tool.

You don't attempt to answer the question of human existence by pursuing truth, writing poetry, or performing scientific experiments. Instead, you make new *things*. Let's make an artificial intelligence; let's make an artificial reality; let's make artificial life to see if we can do it. Nerds actually want to see if they can make new life. They study reality by trying to make a new reality. From this urge has come new territory and insight, which raises other related questions. If we can make reality, what happens when we do this? What happens if we rewire it? What if we change the physics? What if we don't start with carbon to make new life? What if? What if? This is the arena we are rushing into as a culture.

New worlds

This new culture makes tools that create, which throws the spotlight on us as creators with the power to create extremely complex things. Of course, there is an ongoing debate as to whether or not we can create anything as complex as life or a human mind. This remains to be seen, but I think we have to be open to such a possibility, if only because we are getting there so fast. Nerds are on the cover of *Time* magazine; they star in movies about technology itself. The worlds they make, the things they do – this is where we are headed, for better or worse. I'm not saying this is the way it necessarily has to be, but I am saying this is the way we are headed.

Whenever there is a question, nerds try to answer it with technology. They like questions that drive them to create things, even worlds. The world of "Riven," a popular computer game, is made by two brothers who are both Christians; it's a fully immersive world. They are going to make more of these worlds, fully rendered, fully complex, every object, dynamic, as real as you can imagine. These worlds exist in their minds, except that now we too can wander through them. "SimCity" is another virtual world where whole buildings and cities evolve over time. Games like this raise the question of how real a city has to be before it has any "reality" to it.

In a sense, though, this question is mute. Worlds *already* have their own reality. SimCity is a kind of a city, even if it doesn't have the same reality as our lives do. Looking at these worlds stirs up questions about our relationship to these worlds. As their creators, do we have any responsibility toward them? What kind of responsibility do we have? Such questions reflect back into our own world and have the potential to illuminate our views of God.

In the Big Bang model there are three options for the expanding universe. There is the possibility that it expands but doesn't get very big and collapses back upon itself. Another possibility is that it runs off, spreading too thin too quickly for life to develop. What you want is what I call the "Goldilocks point" – not too big, not too small – a universe that keeps going for a very long time. This notion of a Goldilocks point applies directly to these virtual worlds the nerds are creating.

Different ways of being a god

Within the range of possible simulated worlds, nerds are looking for the Goldilocks point. They are looking for that narrow space which allows things in their world to keep building upon themselves. Lots of MUD gods spend their time looking for just the right initial conditions. They spend their time trying various ways of getting their worlds going, and then they just let them go.

The protein DNA – lots of biological systems, in fact – seems to have a Goldilocks point, too. DNA zips and unzips at around body temperature, right around the temperature water exists in its liquid state between gas and ice. The curious thing is that DNA is like a switch in this respect; you can zip it up or unzip it, almost like a transistor switch. Its ability to come apart and come together with small changes in temperature suggests to me that life is computational – it can process information by turning itself off and on, by zipping and unzipping. All the systems the AI community is searching for skate, like the DNA protein, right down the middle toward the Goldilocks point. Like the MUD gods, AI researchers spend lots of time setting up initial conditions and then letting their systems unfold. This is one way of being a god.

But if you were going to make universes and watch them grow more and more complex, what are some other ways you could do that? First, you could be a pyromaniac. Make the biggest bang you can, and let it roll out by itself. This is essentially the approach I described above. Another kind of god is the pilot. This is where you get something going but never quite take your hand off it because you don't want to leave it alone. You start the world going and then you keep making subtle adjustments, nothing severe, just enough to keep things rolling along. This is another way of being a god.

Then there's the interventionist. We are all familiar with this one. You get down into your world when things go a little bit awry and set them back on track. And then there's the improver, the evolver. Perhaps you change your mind, have a new idea about what you want as the world goes along. Of course, there are lots of variations one could think of, but my point is that it might be worth thinking about the different ways in which you can be a god in these universes.

One intriguing thing about the different kinds of god is whether or not they change, and if they do, where this change takes place. Is the god a fixed reality that never changes in any way? Does the god introduce the change into the system? How? All gods have to make decisions about the way change happens in their worlds. And what is it in our world that changes? Science. Science is the religion of change. Even though it can endure over long periods of time, it stresses the idea of change. Might there be something to think about here, too, in relation to our understanding of God?

Let's define God as something that creates something out of nothing. If we go with this definition, then we see that the little gods creating these worlds are doing the same thing. They are creating something out of nothing. Moreover, as a god you can take something from your universe – an object, an entity – and let *it* become a god. Is it clear where this process is headed? God made us, so now we are going to take a part of ourselves and make a new world. Call it SimCity, or Riven, or an artificial intelligence robot. Next, an object from our created world itself makes something, and so it continues all the way down. That's the recursive nature of God. It is not hard to see this as a reflection of God's recursive nature: God was the Creator, therefore we who are made in God's image can create.[1] We can create, not only novels and pennies, but worlds! These worlds may actually have beings with free will, and they themselves may actually create something.

This reminds me of the invention of virtual reality. The fellow who invented it, Jaron Lanier, had goggles and gloves which he could use to make the world and enter it. I was there when he went into his first complete world for the first time. In front of me was a god limiting himself, immersing himself in a world that he created. He discovered things he had no idea were in the world, even though he had made it. It was eery to watch him be so surprised by what he himself had created.

But do these worlds have to be inferior all the way down? We should, I think, be open to another possibility, namely, that we can make things smarter and more powerful than ourselves. This is possible. Granted, we don't know how to do it. But just because we don't know how, doesn't mean we won't be able to with the help of the things we create. We live in I–It and I–Thou worlds, but the I–Robot world is coming.

Information

Let's suppose that, eventually, we will be able to make things smarter and more powerful than ourselves. Some may be animal minds, but others may be different kinds of intelligences. I suspect that we may in fact need other minds to help us further our own picture of God. Other minds we make may be able to see God more clearly than we can.

This whole wired world we are building, this great big rush we are in right now to make technology, is founded on something called "information." In many ways, information is as weird and intangible as the Holy Spirit. When people talk about information, they could just as well be talking about spirits. They wave their hands the same way; they use the same kind of vocabulary. Presently, we don't have a good theory of information. And yet, our whole culture and economy are based on information. Everything in the world is really information. John Wheeler calls it "Its are Bits." Go as far as you can into the cosmos, and you end up with information. Go down as deep as you want, and you find information. It's very much like spirit. Exploring the way that nerds are getting information holds, I think, immense potential for theologians.

Conclusion

As far as we can tell, information is really nothing at all. Structure, numbers – who knows what it is? But it has tremendous power. Lots of people who don't believe in God, believe in information. Isn't it curious? Why do they believe in information? Well … it's there. The power of information is an issue which will come up again and again, as we think about the material, the spiritual, about God. The fact that we are moving into an arena right now where we have the power to create new, long-lived, surprisingly rich worlds surely has something to teach us. As we begin to make minds different from our own, we may be aided by them in understanding who God is and what He wants from us.

Although this essay has been a bit light-hearted, I want to conclude by saying that I take God very seriously. We cannot avoid asking whether the idea of completing God's creation by making our own is a sacred or blasphemous act. Right now, we don't know. I suspect, however, that we will have to tread the line between not letting our efforts become blasphemous, and not letting them become hubris.

Note

1   Mitchell Marcus's essay in this volume expresses similar thoughts about God's recursive nature.

4   Allan Sandage

Allan Sandage is Research Staff Astronomer Emeritus at the Observatories of the Carnegie Institution of Washington. Dr. Sandage's research interests include stellar evolution, observational cosmology, quasars, and galaxy formation and evolution. He is the author of five books and 350 research papers and is currently Associate Editor of *Annual Review of Astronomy and Astrophysics*.

Interview by Philip Clayton

PC: Did you have any sort of early exposure to a religion, and did that affect you as you began your scientific training in the early years?

AS: I really didn't learn anything about Christianity until I was perhaps forty years old. I went to the Methodist church as a young boy, but received no doctrinal training, so it was only through reading about the Christian faith very much later in my life that I gained any understanding of the tradition.

PC: So when you began your scientific training, you didn't have to break from Christian beliefs because you didn't hold any?

AS: I came to spiritual understanding only very much later when I was plagued by the solution to the question of purpose. From age eleven I wanted to be an astronomer. The scientific aspect of the discovery of things in the world was fundamental to me. But from the age of perhaps twenty, I was plagued by the question, "What is the purpose of life?" Trying to find the answer within science led nowhere.

PC: You were driven by a crisis to find a spiritual answer to the question of purpose?

AS: Absolutely. But also by the philosophical question of why there is something instead of nothing. My deep requirement for an answer to those two questions, what's the purpose of life, and what is the explanation of the mystery of existence, led me to study philosophy. I'm also troubled by the problem of understanding the basis of ethics and morality. What is the good? Science is value neutral whereas religion is value intensive.

But I could never go the way of Nietzsche with nihilism, or Camus and Schopenhauer with existentialism. Those were among the saddest people in the world. Each of them died, I think, a very broken human being. I was not impressed with a relative theory of ethics. I think there has to be an absolute, and the only absolute answer is, what is ethical is what God wills. The existence of God can't be proved. That's where faith versus reason comes in. In a way, I am of the same mind as Luther when he said one must reject reason and adopt faith alone, because faith and reason are antipodal to one another. So in some sense I'm schizophrenic.

PC: It almost sounds like a natural theology – there has to be a place for faith because reason doesn't provide these things. Is that how you think of it?

AS: I had a real problem going from my scientific mode to religion, and at one point I came to the conclusion that theology is an experimental subject. The Bible, or Christianity, or any formalized religion set down the laws by which a human being should conduct his or her life, and those laws work.

PC: So there would be a theological science?

AS: I used to say that theology is an experimental science because the laws are set down. Where those laws come from, that is a real question. What is the authenticity of these laws? That's where faith comes in. That's where you must pass a night of great darkness to come to the idea that you can't treat these matters as you would as a reductionistic scientist. And I am a reductionistic scientist.

PC: In what respects are science and religious traditions relevant to each other?

AS: I'm a cosmologist. What I do is study the evolution of the universe. I've been doing that for forty years. A creation event is not, for me, a proof of the existence of God. I would say that no proof is possible.

One can bolster one's faith, I suppose, by saying that there was a creation event. But Big Bang cosmology does not say this is *the creation of the universe*. It is some catastrophic event, but the mystery of creation out of nothing remains a mystery. The Genesis account of creation started science on its way and put to rest all of the previous magic, only to replace it with a deeper magic.

PC: Is overcoming the superstition of earlier mythological religions important for you?

AS: Without monotheism no religion can mean anything.

PC: You seem to be praising Genesis' ability to move beyond a superstitious attitude of the universe.

AS: Yes, although it is replaced by an unexplainable mystery. I cannot understand the mystery of existence without somehow believing in the supernatural. That's a very strange statement for a scientist to make. I'll put it differently. It's only through the supernatural that I can begin to understand the mystery of existence.

PC: In some way, then, your view of the supernatural must leave the universe open to scientific study and knowledge.

AS: The great thing about the universe is that it is basically reasonable. It is not chaotic. That means that it is regular enough to study. There is this

God of philosophers vs. God of scientists

new science of chaos, but that is not how the universe is essentially run. Otherwise, science would be impossible. My belief in the spiritual aspect has come from ethics and morality, but also the question of existence, and the question of design. But all these things point only to a God of the philosophers, not the God of the Scriptures.

PC: How do you bridge that gap in your own thinking or practice?

AS: Both Pascal and Kierkegaard also had this problem of relating the two Gods. They each said there's a great chasm, and you stand on one side, the side of reason, and you look across the chasm to the other side, the side of faith. If you are going to be classically religious and talk about the God of the Scriptures, you have to jump across that chasm.

PC: By focusing on the act of will, you resist people who would say, "Oh, no, I have evidence to move from science or philosophy into religion proper."

AS: I don't think there is any evidence. It is faith, not reason. If there is proof, faith is not needed! But that doesn't mean that I'm enamored with fundamentalist theology. I've studied the question of hermeneutics, and actually went to a fundamentalist church to see exactly what they were saying, and it angered me tremendously because they ridiculed evolution; they say that the earth is only 6,000 years old. In their course on hermeneutics they had the seven laws of interpretation of the Scriptures, and I said well, there has to be an eighth law, that wherever the Scriptures are fundamentally at odds with scientific evidence, then one must change one's hermeneutics of interpretation such that you can accommodate both. I was not one of their better students.

PC: Is the act of will to believe not just that theism is true in general but, let's say, that Jesus Christ was God's son and died and rose for us?

AS: That's jumping across the chasm. For me, that has to be unprovable, but it is a peaceful thought.

PC: Since the report about Jesus's death and resurrection is contained in the Scriptures, then part of the leap is to accept the authority of the Scriptures?

AS: I tell all my scientific friends that I believe the Bible is inerrant. But I can't *really* believe that unless I apply my hermeneutics. The Gnostic interpretation was that his spirit rose, but not his body.

PC: Does a scientist today, a Christian scientist, have to be a Gnostic, because we know that bodies don't get resuscitated?

AS: A spirit may be the thing that rose from the dead. But once I say I'm willing to break the laws of physics, once I accept the supernatural about the creation of matter, the mystery of existence, then I've cooked my goose. Why not let a miracle happen the second or third time?

PC: If God created everything, why couldn't He work miracles inside the world? If He broke the laws of physics once, why not a second time?

AS: Yes, that's what my fundamentalist friends have told me is the answer to my dilemma.

PC: And do you sometimes feel that response as well?

AS: I am content to live with the mystery.

PC: You've given strong reasons for why the separation between science and religion has to be there. Do you view it as schizophrenic, or as how things ought to be?

AS: Science and theology, to me, are completely separate, except to solve the mystery of existence. My theological aspect is a hunger for an absolute basis for living. It has stopped the questioning of purpose, it's made peace in my mind, but has not influenced my science.

PC: Do you feel a tension between these two separate areas: a professional scientist and a practicing Christian, or is their separateness unproblematic?

AS: It's unproblematic because I've come to the conclusion that science cannot answer a great number of questions that come naturally to a thinking, sentient being. Questions of emotion, questions of morality, questions of feeling, of appreciation of music. To say that those are not a part of life is wrong. But science, of and by itself, is the only way to objective truth. We can never reach that kind of truth in theology.

Allan Sandage

Science and religion

Separate closets in the same house

Introduction

When the history of astronomy in the last half of the twentieth century is written, two of its greatest triumphs, stellar evolution and cosmology, will have major seats on the stage. The origins and evolution of structures in the universe (planets, stars, galaxies, clusters of galaxies, and the intrinsic geometry of space itself) have been the recent themes of this development. For nearly fifty years I have been working in astronomy and astrophysics, having joined the Mount Wilson and Palomar Observatories in 1952. When I began my career, astronomers focused largely on what is called "observational astronomy," observing what is out there and calibrating the properties of what we see (absolute luminosities, temperatures, dimensions, distance scales, etc.). But as the field developed from the 1960s to 1980s, the aims throughout astronomy moved from describing what is out there to the problems of "origins" and how these things "got there." Because of this, such inquiries have widely been said by non-scientists to have some connection to religion through such ambiguous words as "creation," "design," and "first cause."

Why did I choose a vocation in astronomy? From an early age I was consumed by the mystery of existence: "Why is there something rather than nothing?" I call this sense of wanting to uncover reality's deepest meaning my "divine discontent." The first moment of astonishment came with the idea that there was a time before the world as we know it existed: there was a time when there were no humans, no living thing, no earth, no solar system, no galaxy, no chemical elements, indeed, no universe. With this realization, perhaps at the age of twelve, "origin research," whatever that may have meant to me at the time, became the dream of a vocation.

Indeed, in the *naïveté* of youth I wondered if the central engine of the universe could be discovered through astronomy. Although the present answer is an emphatic "no," it took much searching to arrive at that answer. In fact, if there is to be any answer to that question, I believe it must lie outside the hard, rationalistic reductionism which is the hallmark of experimental science, both in the laboratory and the observatory.

The central engine of the universe, if this concept has any meaning, has receded for me outside the practice of science. This does not mean that science itself cannot drive one's faith in the realm of human experience. (The mysterious, delicate inter-connections of the co-operative phenomena of a living organism, for example, seem to me to be incapable of arising without a blueprint. How does a mighty oak develop from a tiny acorn? This is what I mean by "science driving faith.") I only mean that there is more to the human view of the world than is contained within the majesty of the cold but exquisite equations of mathematical physics.

Cosmology as science

What is the present scientific situation in cosmology that has such appeal concerning questions of "creation" and "first origins"?

Since the 1950s, astronomers have discovered how stars are born out of the gas of the interstellar medium, burn nuclear fuel and radiate the liberated energy away, thereby shining in the process, subsequently becoming red giants, and end their lives as white dwarfs or black holes, or explode as supernovae. It has also been discovered that there are no stars in our galaxy older than about thirteen billion years – an incomprehensibly large number, but nevertheless a finite number. (A stack of one dollar bills amounting to thirteen billion dollars would reach a height of 200 miles.)

The second triumph in cosmology has been the discovery that the universe as we know it has not existed forever. It also has a datable age that is close to the same age as the oldest stars in our galaxy. The discovery in the 1930s that the universe expands with a linear velocity–distance relation, appearing the same for all observers no matter where they are placed in such a velocity field, was the early primary evidence for a "creation" event, and was itself the birth of Big Bang cosmology.

Since that early understanding in 1929, the evidence supporting the idea of a Big Bang has become overwhelming. In addition to the initial evidence that the universe is expanding with the required linear velocity–distance law, newer evidence includes (i) the discovery that all galaxies are nearly the same age of fifteen billion years, (ii) the chemical elements have not existed forever – their initial appearance has also been dated at between ten and twenty billion years based on the implicit time scale in radioactive decay, (iii) discovery of the 3° K background radiation that has a precise Planckian spectrum, explained as the left-over radiation of a hot initial plasma that has cooled by the expansion, and (iv) the discovery of minute variations in the temperature of the background radiation (by the COBE satellite experiment) that are required (the Sacks–Wolfe effect) to make galaxies out of the Big Bang event by gravitational instability.

The most persuasive argument for the Big Bang, before the discovery of the relic radiation, is the remarkable agreement of the three time scales of (i) the age of the oldest stars based on the astrophysics of stellar evolution, (ii) the expansion rate that dates the "creation" event, and (iii) the age of the chemical elements.

Cosmology is not religion

Much has been made by non-scientists, often by theologians and other religious commentators, that Big Bang cosmology supports any number of religious ideas, from the existence of God, to claims that the Bible is a book of science because of a supposed connection between scientific cosmology and the first chapter of Genesis.

Scientific cosmology is not a religion. It has nothing to say about the mystery of existence itself. As a branch of physics, the subject only concerns the unfolding process of the universe once the Big Bang occurred, for whatever the reason; and that "reason" is outside the purview of science. Furthermore, knowledge of the "creation" event gives no knowledge of "the Creator," or why such an event occurred.

Nevertheless, the so-called cosmological argument concerning the "first cause" of the universe was one of the classical (ecclesiastical) proofs of the existence of God. This "proof" has been discussed since the time of Augustine, predating the advent of modern scientific cosmology by 1,600 years. This argument is not scientific, but through scientific cosmology we come perhaps as close to theology as science can come and still remain science. Because the "creation" event is outside science, it is only through the supernatural that the mystery of existence can be understood.

This answer to my youthful divine discontent came relatively late in the game, when I finally willed myself to believe. Although this went against my training as a reductionist scientist (one thinks, for example, of the extreme scientific creed given by William Clifford (1970): "It is wrong always, everywhere, and for anyone, to believe anything upon insufficient evidence"), I have since learned that very little of what each of us believes, both inside and outside of science, in fact has sufficient evidence. The judgment of what counts as sufficient of course varies over the entire spectrum from atheist to fundamentalist.

What does "creation" mean?

The word "creation" is sometimes seen as inflammatory rhetoric by those who insist that Big Bang cosmology says nothing about the existence of space, time, or matter before the Big Bang. Was this event a creation *ex nihilo* (creation out of nothing), or only a change (a drastic one, to be sure) from a previous state to one out of which the world as we know it developed? When someone objected to his use of the word "creation" in a lecture, George Gamow replied: "If you really believe that I use the word incorrectly, why not simply assume that we mean 'creation' in the same sense that people talk of a woman's fashion as a 'creation'?" This, then, is the only sense in which science can talk about creation. Science cannot get to the uncaused "first cause," which is outside of physics.

I do not believe that the universe can create itself. Typical attempts to provide a scientific answer to the question "Can the universe create itself?" implicitly start with a "first" universe, out of which other universes are generated. It is here that I would ask Andrei Linde where his first universe came from that spawns his multiple baby universes *ad infinitum*.

Genesis

Does scientific cosmology conflict with the book of Genesis in the Bible? No, not with regard to the issue of creation.

Herman Wouk in his book *This is My God* says that the creation myth in the first chapter of Genesis, adapted from the Babylonian creation myth itself, made science possible because with one stroke it did away with the "charming" Greek and Roman gods of one thing and another, replacing them with a single Force that created the vast machinery of the world and set it going. Wouk writes:

> The first chapter of Genesis cuts through the murk of ancient mythology. THE universe is proclaimed a natural order created and unfolded as set going by one force like a vast machine to proceed under its own power. There were no man-like gods. No sun god, no moon god, no love god, no sea god, no war god. At last even the charming Greek and Roman gods withered under the stroke. Genesis is the dividing line between contemporary intelligence and primitive muddle.
>
> (Wouk 1974: 36)

Regarding the apparent time-scale discrepancy, the allegorical creation myth of Genesis says nothing about a literal six-day creation in terms of days as we know them *now*. But isn't it unfair to see the first chapter of Genesis as an allegory? I don't think so. Elsewhere the Bible cautions us against taking itself literally: "But do not let this one fact escape your notice that with the Lord one day is as a thousand years, and a thousand years is as one day" (2 Peter 3:8); or "For a thousand years in your sight are like yesterday" (Psalm 90:4).

Yet "creation" is nonetheless outside the realm of science. To say that the universe was created out of nothing by a "quantum fluctuation in the false vacuum" is simply promiscuous use of smoke and mirrors; it is empty speculation with no experimental basis except that the universe does exist. Scientists who say that "we do not need to invoke a higher power" simply mandate away any possible explanation outside the exceedingly narrow precepts of reductionist science. (What indeed is the origin of the false vacuum?)

Of course, *ab initio* creation itself is, by any definition of the boundaries of science, supernatural. To put all the onus of creation on an unknown creator is seen by atheists as the height of absurdity; but for others, the alternative of believing that the universe created itself out of nothing is even more absurd. Chesterton writes of the choice in his very clever book *Orthodoxy*:

> The whole secret of mysticism is this: that a man can understand every-thing by what he does not understand. The scientist seeks to make everything lucid (by denying miracles), yet succeeds in making every-thing mysterious. The mystic allows one thing to be mysterious (permits one miracle), and everything else becomes lucid.
>
> (Chesterton 1959: 28)

It comes down to the necessity to choose. The atheists, agnostics, and "almost believers" refuse to choose because they are waiting for what they cannot have – proof! Luther said it right. No proof is possible. No proof is necessary.

Need science and religion be in conflict?

There are proper boundaries for both science and religion, each of which discusses different aspects of existence. Physical science and molecular biology discuss the world studied by scientists as they attempt to step outside themselves and remove themselves from the inquiry. They pretend they are not part of the world they investigate, and consequently believe that in doing so they can find objective truth. It would seem that lower animals have no such "external" sense of themselves with which to analyze the universe in a "disinterested" sense. But this sense comes with a certain cost. As Frank says in one episode of the comic strip *Frank and Ernest*, "I stepped back to put things in perspective and before I knew it I was out of the picture altogether." It may be impossible to draw an impenetrable line between objective and subjective truth.

With Augustine, Abelard, Anselm, and many others, I believe that there cannot be conflict between science and religion if we get the hermeneutics right. *The science and methodology of interpretation*

> *The Bible*: Now faith is the assurance of things hoped for, the conviction of things not seen.
>
> (Hebrews 11:1)

> *Augustine*: Faith precedes understanding.

> *Abelard*: Science and religion cannot conflict within their own realm. Otherwise, God who made them both would be in contradiction.

> *Anselm*: I do not seek understanding in order to believe. Rather I believe in order to understand.

The necessity for religion

Science can address only a very limited range of problems. Most of life's problems lie outside the realm of science and reside in ourselves as subjective subjects. Spirituality is one of these areas.

John Polkinghorne writes in his *The Faith of a Physicist*:

> The great success of science is purchased through the modesty of its ambition, restricting the phenomena it is prepared to discuss to those of an impersonal [objective rather than subjective] and largely repeatable character.
>
> (Polkinghorne 1996: 5)

Because science is *value neutral*, and religion is *value intensive*, religion is a necessity for man as a spiritual being. The inability of science to provide a basis for existence, purpose, moral value, and free will is what makes religion necessary. Karen Armstrong (1993: xx) in her book *A History of God* writes, "Even if God does not exist, He still is the most important reality in the world."

The character of science and religion

Both science and religion proceed by making models and testing them, discarding those parts of the models that do not work and refining those parts that do.

Differences in testing the models of each lie in the differences between objective (scientific) and subjective (religious) truth. Our daily lives are flooded with subjective truths. Our actions depend on how we react to our feelings. Although theology is concerned with experience, and thus in a sense can be tested by experience, no *objective proof* seems possible in theology in the way that science provides proof. But, believers can say with Luther, "no proof is necessary." They say that religion works for them and is therefore true. This idea, which is Kierkegaard's view of subjective truth, is more important in everyday life for most humans than the abstract models of theoretical physics. Physicists, by and large, are Platonists who seek reality in the archetypes behind the scenes. Non-scientists, by and large, are Kierkegaardeans for whom the subjectivity of life and thought is more real than scientific models.

In his well-known book *God and the Astronomers*, Robert Jastrow writes:

> The scientist has scaled the mountains of ignorance; he is about to conquer the highest peak; as he pulls himself over the final rock, he is greeted by a band of theologians who have been sitting there for centuries.
>
> (Jastrow 1978: 116)

My view is somewhat different. I believe that the realms of science and religion are nearly orthogonal. It seems to me that scientists and theologians have climbed closely adjacent but different peaks. When each has reached their separate summits they can view one another, even exchange arguments and claims of hegemony over one another, but they are not close enough to one another for either to play king of the other's mountain.

How to reconcile the God of the philosophers with the God of the Scriptures

It is one thing to believe that an organizing Force is required as the "first cause" in the "creation" of the universe, but it is quite another to identify this Force with the God of the Scriptures. In moving from one to the other, one begins the journey by contemplating the deep canyon (the abyss) and the high plateau of reason on the one side and that of faith on the other. One cannot wait on the side of reason until one is certain. You must choose! Kierkegaard chose to leap blindly *into* the abyss, from the side of reason without knowing if his spring was powerful enough to propel him to the other side. His leap, being blind, relied on the hope that his faith was correct. Pascal, on the other hand, leapt with the certainty of true faith. My approach has been to decide to believe in the Scriptures as an allegory for the great central engine of the universe behind the scenes, explaining what science cannot explain – namely *existence*, *purpose*, *value* and *free will*.

The fruits of faith as a pragmatic solution to the divine discontent of my youth with the deepest of problems of existence have not been insignificant. Faith has provided a brace against nihilism and the tragic despair of Nietzsche and Schopenhauer; an explanation of the problem of existence through belief in a Creator, as absurd as that belief may be (on his death bed, the great nihilist of the existential movement, Jean-Paul Sartre, called for a Catholic priest to give him last rites, "just in case"); and a theology which, at some level, explains consciousness.

I leave the reader to ponder several ways in which science itself remains mysterious: Why does mathematics so effectively describe the world? Why does Newtonian gravity, which depends upon the mysterious, even absurd notion of action at a distance, enable us to send satellites to the moon? The curvature of space as an explanation of gravity (the equivalence of inertial and gravitational mass) is even more mysterious. The miracle of Maxwell's equations from which come electromagnetism, light, radio waves, and communication over large distances with no contact "forces," is an even greater mystery. And finally, modern science gives us the strange, counter-intuitive world of quantum mechanics. As Niels Bohr said, "Whoever says they understand quantum mechanics does not understand quantum mechanics."

In the end, both science and religion build models as guides to understand the mysterious reality behind the scenes. Models are all we have, both in science and in religion.

References

Armstrong, Karen (1993) *A History of God*, New York: Ballantine Books.

Chesterton, G.K. (1959) *Orthodoxy*, New York: Doubleday/Image Books.

Clifford, William K. (1970) "The Ethics of Belief" in *The Rationality of Belief in God*, edited by George I. Mavrodes, Englewood Cliffs, N.J.: Prentice Hall. Originally published in William K. Clifford, *Lectures and Essays*, London: Macmillan and Co., 1886.

Jastrow, Robert (1978) *God and the Astronomers*, New York: W.W. Norton.

Polkinghorne, John (1996) *The Faith of a Physicist*, Minneapolis: Fortress Press.

Wouk, Herman (1974) *This is My God*, New York: Simon & Schuster.

☆ Science and faith are not mutually exclusive; nor can the twain become one; there is an inherent and inevitable tension between the two forces,

- both seeks answers to important questions.
- both lead to fundamentalist and radical believers who worship their own creations

Ideally

Science begins with a premise, and seeks truth as the answer

faith begins with a truth, and seeks a premise as an answer

much like that between water between a liquid and a vapour

5   Cyril Domb

Cyril Domb is Emeritus Professor of Physics at Bar-Ilan University in Israel. Before settling in Israel he served from 1954 to 1981 as Maxwell Professor of Theoretical Physics at King's College, London. Since 1950, Professor Domb has published numerous articles in the area of science and religion which have appeared in a variety of books and journals in different countries. Dr. Domb currently edits *B.D.D.*, a journal published by Bar-Ilan University with articles in Hebrew and English in the area of Torah and scholarship. He is a Fellow of the Royal Society.

Interview by Philip Clayton

PC: I would like to start by asking about your religious upbringing and practice and the influence that's had on you.

CD: I was born in England to a very traditional Orthodox Jewish family. My mother and father were both born in Poland. My parents were committed Orthodox Jews, but they had to adjust to the mores of British society. My grandparents never did, and I was more or less brought up in my grandparents' household. I think that my very strong religious commitment came from the home which they ran and from the ideas which they put forward.

PC: Was there any influence of your early religious practice on your decision to become a scientist?

CD: Not really. I was good at mathematics at an early age. I was very lucky that I had dedicated teachers. They'd give me more advanced books and so on. The really crucial thing for me was when the teacher at my high school decided to put me in for a Cambridge scholarship. I was seventeen, and the normal age is eighteen or nineteen. I never dreamt that I would pull it off, but I did, and I went to Cambridge – it was like being taken, from the Jewish religious point of view, from a hothouse, and thrown into cold water. It was a challenge to maintaining my Judaism, and an intellectual challenge. I had to face up to both of these, and it was very tough. But I think that this did me a world of good, because it gave me confidence ever afterwards. If I got through that okay, then I could face any other challenge.

PC: As you look back over your career as a scientist, in what respects do you see your Jewish practice and belief and your scientific work as being relevant?

CD: Well, at present they have evolved to being very closely linked. They may not have been at the beginning. The fundamental problem, in its most primitive form, is that science seems to know everything. If everything is governed by scientific laws, where does God come in?

I was very fortunate that I heard an extremely good lecture after the war on science and religion by Professor Adolf Frenkel. He gave a talk on science and religion in which he explained exactly how science changes, how science is tentative. I learned for the first time that these great scientific theories that one heard so much about really are not absolute truth, that they're only tentative summaries of our situation, whereas religion deals with eternal values, with what is right and what is wrong, and with many of the major driving forces in one's life.

When people would ask me, "What is more important in your life, your science or your religion?" I always used to say to them, "If somebody would give me a million dollars to abandon science, I would accept it. If somebody would give me a million dollars to break the Sabbath, then I won't." It's quite clear that my commitment to religion is a far greater commitment than my commitment to science.

PC: So science is a logical, revisable attempt to understand the world, and religion is a fundamental practice of belief, values, life-style, commitment?

CD: Yes, except that one also uses logic and reason in religion. Religion requires commitment, but it's not totally that way. I have commitment to certain things: to a Creator who is concerned with the universe, to free will, and so on. But once you've made your assumptions you don't abandon reason. I was brought up by my grandparents studying Talmud, where logic is absolutely the driving force. It certainly had that in common with my mathematics and science. Science and religion are only different in the sense that scientific truth is not absolute. Some new experimental fact can come along and the whole theory will be overturned. You can't do that with religion. Religious truth is absolute.

PC: The next area I would like to discuss involves a belief shared among Judaism, Christianity and Islam, that the universe has been created by a personal being who has a particular purpose in the act of creation and a destiny for the universe as a whole. It's more than just an arrow of time. It's also the belief that there was design, and that there is a *telos* to the universe.

CD: Yes, there was a design and purpose. Purpose is very difficult for us to establish, except by looking at history. But design one can certainly establish. When I started, the religious person was on the defensive. The atheist was always attacking. Now it seems to me the boot is on the other foot with all these incredible discoveries: the fact that radiation is created in such a way, that you can know what is the constitution of a planet billions of miles away, the fact that you can measure enormous distances, the fact that we can measure the size of the universe ... from the point of view of plausibility, that all this fascinating complex structure could just have happened on its own would be an incredible coincidence.

PC: In other words, theistic belief was viewed as implausible when you started doing physics and is now not only plausible, but somehow supported by the developments in physics?

CD: For the person who wants to accept the support, yes, this is the transformation that's happened during my scientific career. The human person is a being with moral responsibility, with freedom of choice, with a special capacity for relationship with God.

PC: How does that religious notion of person fit with contemporary theory in the sciences? Is it something that we can derive, or is it a major break?

CD: Very little challenge comes to this from the physical sciences, although a challenge did come, of course, on the question of free will. There was a challenge from the physical sciences, the famous Laplacian challenge. Laplace more or less said, "Tell me exactly what the state of the universe is at one stage, and if I'm clever enough, I will be able to predict it ever afterwards." That was overturned with the advent of the quantum theory. It's very satisfying that Laplace has been overturned because it teaches that science never deals with certainty. But our belief in free will is much deeper than that. If man doesn't have free will, then religion is meaningless. Without it, you can't improve people, you can't try and make them behave better.

I think the major force against religion and religious values in the past century has been from biology, from Darwinism. I don't think that physics has presented anything like such a challenge. In fact, physics has provided the anthropic principle, which provides clear evidence of design, that man didn't just emerge by chance, that the whole universe seems to have been arranged so that man could survive and emerge. The anthropic principle has its opponents, but many of its supporters are people who, when they started their science, had no inclination towards design or religion.

Now, of course, a religious person doesn't start sitting on the fence. If you're going to say, "Well, if the science works out, I will believe, and if the science doesn't work out, I won't believe," that's not religion. I used to think once upon a time that doing science, having a wonderful brain, and seeing marvelous things and pioneering in this way, would make one a more moral person, but it doesn't.

PC: What are the fundamental developments in physics in this century that changed the climate and made theism so much more palatable?

CD: Partly an appreciation of the transience of scientific theories, the overturning of Newton, the overturning of Laplace, the strangeness of the quantum theory, the discovery of DNA, all of these tremendous discoveries. The door is still open for the people who want to dig their heels in and say, "I don't accept this." But the ordinary scientists have seen all these things and I think it gave them a gut feeling that something is behind it all.

PC: In terms of this response to science, that there must be a design behind the universe as we observe it, do you sense a fundamental affinity among theists, whether Christian, Jew or Muslim, in their way of seeing the physical universe as expressing the nature and activity of a creator? Is this a point where those three religious traditions, in your view, draw together?

CD: They should, I don't know whether they actually do. I can only think of individuals. I don't know much about the Muslim world, but the person who represented this for me more than anybody else was Abdus Salaam. When he won his Nobel prize he went to the mosque to give thanks to the Almighty for having done this for him. I sent him a telegram congratulating him and saying that I'd heard this and how wonderful it was in our own age to have people like him. In the nineteenth century many of the great scientists were deeply religious. I told him I was happy to find the religious scientific tradition of Faraday and Maxwell being maintained in our own age. A couple of weeks later, I got a reply saying, "There are few letters which have given me such pleasure. Bless you for writing. May I quote your letter sometimes?"

Cyril Domb

Does science offer evidence of a purpose and a transcendent reality?

Introduction

Religion starts with a commitment to a number of basic beliefs. The details of these beliefs vary from faith to faith, and I do not think that they can be subjected to experimental test. They may conflict with current scientific ideas and thinking, in which case they merit re-examination. The classic conflict between geocentricity and heliocentricity led most serious religious thinkers to the conclusion that geocentricity was not an essential item of belief. But the determinism of nineteenth-century Newtonian mechanics with its negation of divine providence and free will could not be accepted. This time it was science which had to retract; twentieth-century theories have abandoned the deterministic picture and talk instead in terms of probability. These examples indicate that the interaction between science and religion exercises a healthy influence on both.

I shall consider essential items of belief shared by the monotheistic faiths coming under the heading of "purpose" and "transcendent reality." My thesis will be that although these religious commitments are not based on science, a religious scientist can derive inspiration and support for them from scientific developments in the past few decades. I shall suggest that these basic tenets of the monotheistic faiths are compatible with modern scientific thinking, and that a harmonious synthesis can be achieved now more than at any previous time in history.

I begin with the striking reassessment of the status of scientific ideas by the secular[1] philosopher of science, Sir Karl Popper.

> The old scientific ideal of absolute certain demonstrable knowledge has proved to be an idol. The demand for scientific objectivity makes it inevitable that every scientific statement must remain tentative forever. Only in our subjective experiences of conviction, in our subjective faith, can we be "absolutely certain."
>
> (Popper 1959: 280)

It is this subjective faith with which I shall concern myself in this essay.

Existence of a Creator

Among the early pioneers in astronomy and physics who felt that the laws of nature revealed the remarkable handiwork of the Creator were Johannes Kepler, Robert Boyle, Isaac Newton, Michael Faraday and James Clerk Maxwell (Jaki 1966: chapter 10). But they were men of religious belief and practice. A new phenomenon in this century is the affirmation of the Creator by *secular* scientists. Here are three typical examples.

The first and most notable is Albert Einstein. The following passage from the autobiography of his collaborator Leopold Infeld summarizes Einstein's outlook:

> Nothing is as important as physics. No human relations, no personal life, are as essential as thought and the comprehension of how "God created the world." In this phrase so often repeated by Einstein with variations, was his peculiar religious feeling that laws of nature can be formulated simply and beautifully. When he had a new idea he asked himself: "Could God have created the world in this way?" or "Is this mathematical structure worthy of God?"
>
> (Infeld 1980: 16)

In 1970 Freeman Dyson spoke at the opening of a new physics laboratory at Princeton University, wishing to emphasize the importance of astrophysics and biophysics as worthy of serious attention by the new generation of research physicists. The following quotation is taken from his address:

> I have heard some accelerator enthusiasts talk as if they seriously expect by building one more machine and measuring a few cross sections to solve all the outstanding riddles of nature. I do not believe that anybody can read God's mind as easily as that. ... I would be disappointed and I would consider that the Creator had been uncharacteristically lacking in imagination if it turned out that no surprises remained in the vast range of energies beyond the reach of accelerators.
>
> (Dyson 1970: 26)

The final example relates to my personal experience with Fred Hoyle, a pioneer of astrophysics and cosmology in the twentieth century. I first met Fred in 1941 when I joined the British Admiralty's Radar Research Group after graduating from Cambridge University. He was some years senior to me and had been a Junior Fellow of St. John's College; although he was then a confirmed atheist whilst I was an observant Jew, we established a rapport and friendship which has continued until the present day.

Fred had been married a year or two before, and for sentimental reasons his young bride had asked for a church wedding. Fred acceded to her request, but took the vicar aside to point out that he was an atheist; the latter responded considerately that he would make the ceremony as short as possible.

In 1994 Fred published his autobiography *Home is Where the Wind Blows*. On the final page he sums up as follows:

> The atheistic view that the Universe just happens to be here without purpose and yet with exquisite logical structure appears to me to be obtuse. ... One can conceive of various universes defined by different forms of mathematical restrictions. What I suspect is that the restrictions defining "our universe" are not just any old restrictions. The restrictions are optimized for their consequences. Or, to put it another way, God is doing His best, and to load off onto Him the all-powerful concept is a gross insult, an insult by people who do not merit the great trouble which has been taken on their behalf.
>
> (Hoyle 1994: 421)

This may not be conventional theology, but it is at least a theology.

Divine providence

> It seems to me that if the word 'God' is to be of any use, it should be taken to mean an interested God, a creator and lawgiver who has established not only the laws of nature and the universe but also standards of good and evil, some personality that is concerned with our actions, something in short that it is appropriate for us to worship. This is the God that has mattered to men and women throughout history.
>
> (Weinberg 1993: 195)

Remarkably enough this passage is also taken from the perceptive analysis of a secular scientist, Steven Weinberg, in which he upbraids his colleagues (including Einstein) for believing in a God who reveals himself in the orderly harmony of nature, but does not concern himself with the fates and actions of human beings. Weinberg suggests that this makes the concept of God not so much wrong as unimportant.

This, of course, is precisely the claim of the monotheistic faiths. When God introduces Himself to Israel at Mount Sinai to give the Ten Commandments, he does not use the description "Creator of the Universe." The first Commandment (Exodus 20:2) reads "I am the Lord your God who brought you out of the land of Egypt," which emphasizes His direct concern with human affairs. The Old Testament contains detailed descriptions of how, on important occasions, God satisfies human needs. He causes a very powerful wind to spring up and split the Red Sea; He directs huge flocks of quails to the Israelites in the desert to satisfy their demand for meat; He answers the prayers of Hannah, who was barren, and she gives birth to a male child; when the

Assyrian hordes of Sennacherib besiege Jerusalem and King Hezekiah prays for divine help, He causes pestilence to break out among the besieging army so that they hastily make tracks for home. There are hundreds of other occurrences in the Old Testament characterized as the work of God. I have taken my examples from Jewish sacred literature, but I am sure that my colleagues of other faiths could find appropriate quotations from theirs as well.

Many of these incidents are referred to as miracles. Do they involve contravention of the laws of nature? Is it possible for laws of nature to be contravened? The traditional Jewish view is that God who created the laws of nature has the power to revoke them but does so only on very rare occasions.

What is the character of the many other "miraculous" occurrences recorded in the pages of the Old Testament? Nachmanides asserts that many Old Testament miracles are unusual events which take place after having been previously predicted. The migration of quails is a well-known natural phenomenon; the miracle arises in their being at hand exactly when they were required, as God had foretold to Moses. Epidemics are a feature of everyday life. For an epidemic to break out in the Assyrian army so that it is forced to abandon the siege of Jerusalem in conformity with the prediction of the prophet Isaiah is a miracle.

Many of the occurrences are thus within the framework of the laws of nature, particularly current laws which allow for the possibility of extraordinary events with extremely low probability. But some are definitely outside the framework of natural laws; do they not conflict with science? In fact they do not because they are unique and non-repeatable, and science concerns itself only with repeatable events.

Those committed to religion assume that God is interacting with the world at the present time, and they pray to him to fulfill their needs. For example, prayers for rain are part of the regular Jewish daily service; a special procedure exists for offering prayers for the sick to be restored to good health; and since the time of the patriarchs childless couples have prayed that they may be granted fertility. The belief that such prayers can be answered conflicts with the deterministic universe of the nineteenth century, but we have seen that this description has been discredited. Chance and randomness are ubiquitous in the modern analysis of complex systems, and indeterminacy is prevalent. But the religious adherent assumes that what is indeterminate for human beings is known to God who has access to all the knowledge and power which is logically available.

The example of rain provides a particularly interesting illustration since until recently it was assumed that meteorological phenomena are deterministic, but so complicated that we have not learned to unravel them. As we have already mentioned the prevalent current view leans heavily toward unpredictability (Lorenz 1993: 102–103). There is little difficulty in finding plausible random incidents by which the biblical quails could be directed from their normal migratory path, or by which suitable bacteria or viruses could initiate an epidemic among the Assyrians.

Despite the fact that the tools for divine interaction with the world seem to be available within the framework of the laws of nature, one must recognize that only a small percentage of scientists accept a divine role in human affairs. The recognition of such a role requires a combination of discernment and belief, and even unusual combinations of events can be explained away as "coincidences" or "good luck." This is the attitude taken by most secular scientists.

Design and purpose: the anthropic principle

Scientific opposition to design and purpose in the universe stems mainly from proponents of the theory of evolution.

> Man is the result of a purposeless and materialistic process that did not have him in mind. He was not planned. He is a state of matter, a form of life, a sort of animal, and a species of the order primates, akin nearly or remotely to all of life and indeed to all that is material.
>
> (Simpson 1967: 345)

This dogmatic statement is taken from the epilogue to *The Meaning of Evolution* by George Gaylord Simpson, one of the founders of the modern synthetic theory of evolution. Comparable statements can be found in the writings of Jacques Monod or Richard Dawkins.

The publication in 1859 of Darwin's *Origin of Species* came as a shock to religious thinkers. Darwin suggested that the origin of different species is to be sought not in sharp and discontinuous processes of creation, but in gradual changes accumulating over enormous periods of time. The changes are induced by a process of natural selection, the species which survive being those most fitted to the environment. Hence there is a continuous chain of different species between lower and higher animals, and between higher animals and man. Darwin's view differed radically from the account of creation given in Genesis which had held sway previously.

When the shock had subsided, it was realized that adjustments might be made to existing beliefs which could achieve a *modus vivendi* with Darwinian theory. The "day" in the biblical account can be interpreted as a period of time, and one can resort to metaphorical interpretation of a sacred text when a conflict arises with established scientific results.

But although natural selection should be accepted as an established scientific fact, its function in nature must be regarded as a means for achieving a divine objective. No serious religious thinker can come to terms with a mechanism which is merely random, purposeless and materialistic.

For a century after the publication of Darwin's classic text there were virtually no *scientific* challenges to the theory. In the present century evolution received a great boost from molecular biology, and it was possible to suggest the actual molecular processes involved in the operation of natural selection. Some scientists claimed that a complete chain could be identified down to the lowest forms of living organism, and hence that the origin of life could be explained.

During the last four decades, however, serious and well-founded criticisms of the theory have made their appearance. Darwin's idea of small gradual changes does not fit the fossil record, and the alternative *punctuated equilibria* theory of Niles Eldredge and Stephen Gould has been advanced to account for the sudden appearance and disappearance of species. A totally new mechanism for the extinction of species not envisaged by Darwin has been proposed by a physicist, Luis W. Alvarez (1987) – the impact of meteors or comets colliding with the earth. This "impact theory" has been outstandingly successful in accounting for the disappearance of dinosaurs, and fifteen different predictions of the theory have been confirmed. The theory is now widely accepted by paleontologists; as a result, "survival of the fittest" must now be joined by "survival of the luckiest" (Raup 1991).

Moreover, the theory of evolution cannot be entirely reconciled with the new discoveries of cell biochemistry (Behe 1996); and the randomness of natural selection postulated by so many of the proponents of evolution clashes with the requirements of information theory (Spetner 1997). Above all, the newly formulated anthropic principle shows clear evidence of purpose and planning in the universe.

John Barrow is co-author of a comprehensive treatise on the anthropic principle (Barrow and Tipler 1986) in which he provides striking evidence of fine-tuning in the structure of the universe. Considerations of this kind led Freeman Dyson to declare:

> I do not feel like an alien in the universe. The more I examine the universe and study the details of its architecture, the more evidence I find that the universe in some sense must have known that we were coming.
>
> (Dyson 1979: 250)

I would like to return briefly to some remarkable coincidences (Aviezer 1997) in the theory just referred to of Luis Alvarez. In discussing the implications of his impact theory he emphasized the intimate connection between human beings and dinosaurs in the following words:

> From our human point of view, that impact was one of the most important single events in the history of our planet. Had it not taken place, the largest mammals alive today might still resemble the rat-like creatures that were then scurrying around trying to avoid being devoured by dinosaurs.
>
> (Alvarez 1987: 30)

But there is more to the story. For human beings to exist today, it was not sufficient merely that such an impact with the meteor occurred. The impact had to have occurred with *just the right strength*. As Alvarez explains:

If the impact had been weaker, no species would have become extinct; the mammals would still be subordinate to the dinosaurs, and I couldn't be writing this article. If the impact had been stronger, all life on this planet would have ceased, and again, I wouldn't be writing this article. That tells me that the impact must have been of just the right strength [to ensure that] the mammals survived, while the dinosaurs didn't.

(Alvarez 1987: 33)

It has recently become clear to scientists that the sudden destruction of all the world's dinosaurs is just one of a long series of completely unexpected, highly improbable events whose occurrence was necessary for the existence of human beings – and *all these events just happened to occur in precisely the required sequence.* Indeed this is a major theme in the recent book, entitled *Wonderful Life,* by Stephen Jay Gould, one of the world's leading authorities on evolutionary biology. Again and again, Gould emphasizes how amazing it is that human beings exist at all, because we are an improbable and fragile entity, the result of a "staggeringly improbable series of events ... utterly unpredictable and quite unrepeatable" (Gould 1989: 14). His book abounds with examples of the anthropic principle, leading him to conclude:

It fills us with a kind of amazement [because of its improbability] that humans ever evolved at all. Replay the tape a million times from [the same] beginning, and I doubt that anything like *Homo sapiens* would ever appear again. It is, indeed, a wonderful life.

(Gould 1989: 289)

The phenomena described above represent a challenge to the secular scientists. They can be dismissed as merely an unusual combination of rare events, but the religious explanation that they are a manifestation of divine planning and purpose is surely more plausible.

Free will

I would like now to touch on the question of whether purpose in the universe extends to human beings. If so, it must be assumed that their behavior is not pre-determined, and can be influenced by teaching and example. Indeed, the existence of free will and personal responsibility for one's actions is a cardinal tenet of the monotheistic faiths. But clearly genetic endowment and environmental influences do have relevance to moral behavior. In order to sort out the interaction between moral advancement and genetic and environmental influences, an attempt must be made to define more precisely what is meant by *free will.*

First it must be emphasized that many of the choices which a person faces in everyday life *are* pre-determined, and could be predicted from a detailed knowledge of his character and make-up, or in modern terminology, "how he is programmed." To take a mundane example, when we ask our friends "Tea or coffee?" it seems likely that the choice is not an exercise in free will; for some we know the answer will always be "coffee," while for others it may depend on what they have been eating or drinking during the previous few hours.

However, religion concerns itself with *moral* choices and here the views of an original Jewish thinker of the past century, Rabbi E.L. Dessler, on this topic are very enlightening (1954). Rabbi Dessler first points out that we do not demand that every person is free to make or reject the right moral choice in all circumstances. We are all familiar with individuals about whom we can testify that they would never commit the basic forms of wrongdoing like stealing or murder. There are others of whom we could be convinced that in certain circumstances they could not withstand certain immoral temptations. We must construct a moral ladder with an indefinite number of rungs, in which the lowest rungs represent the crudest types of moral challenge; as we move up the ladder the challenges become more delicate and refined. A person, for whom crude theft represented by a low rung is unthinkable, might have few qualms about a higher rung which may represent making an unrealistic insurance claim, or taking an excessive profit in a business transaction. For him, although the lower rungs are closed, there is a range of open behavior grouped around the higher rung.

For each individual at each stage of his life there exists a point on the ladder identifying a range within which free will operates. The purpose of moral and religious training is to move this point upwards. The point at which a particular individual *starts* his moral career will depend on his genetic endowment and environmental background. His moral achievement during any period of his life is not measured by his absolute position on the ladder, but by the amount by which he has progressed.[2]

Most secular scientists nowadays accept the principle of free will, and although there is opposition from genetics and psychology, it may arise from a failure to grasp the requirements of the model of free will outlined above.

A striking example of the triumph of the exercise of free will is represented by Stephen Hawking. Twenty years ago he was told by the best science of the time that the disease from which he was suffering would kill him within two years. But he refused to be intimidated, and as the disease progressed, he mobilized the best modern technology to enable him to carry on. In the past few years he has become something of a legend. Whenever he lectures, the hall is filled to capacity, and his book *A Brief History of Time* has sold many millions of copies. His lectures and books are good, but not significantly better than those of many other colleagues in his field. It is the triumph of the human spirit over adversity which has attracted such wide publicity and acclaim.

Rationale of the religious scientist

What is the origin of religious commitment? The reasons which motivate people to remain attached to a particular faith with complete conviction are usually complex and personal: family, upbringing, education, history, inspirational literature, personal experience, and perhaps most of all, the need for values which are lacking in science. Religion tries to make people into better human beings, and I think religious scientists would agree that it is more important to be a model human being than a model scientist.

Science plays little part in *initiating* the commitment, and on a number of occasions in the past, as I have indicated, religious commitment has conflicted with current scientific ideas. It is my contention that as a result of scientific progress in the past few decades, all major sources of conflict have been removed. An harmonious synthesis is now possible in which a religious scientist derives inspiration and encouragement from the incredible progress in science in the last century.

Einstein once remarked that the most unintelligible thing about the universe was its intelligibility. He was referring to the fact that a small number of hypotheses are capable of explaining masses of experimental observations. But no less surprising are the data themselves, relating to features where there could be no *a priori* expectation that measurement would be possible. It is as if the Creator deliberately provided the keys through which it is possible for human beings to assess the extent and magnificence of His creation.

Steven Weinberg's observation that "the more the universe seems comprehensible, the more it seems pointless" (Weinberg 1993: 204) – a view shared by quite a number of his fellow cosmologists and physicists – is the antithesis of how the religious scientist sees the universe. The latter identifies instead with the outlook of another secular physicist, Richard Feynman:

> The same thrill, the same awe and mystery, come again and again when we look at any problem deeply enough. With more knowledge comes deeper more wonderful mystery, luring one on to penetrate deeper still. Never concerned that the answer may prove disappointing, but with pleasure and confidence we turn over each new stone to find unimagined strangeness leading on to more wonderful questions and mysteries – certainly a grand adventure!
>
>   It is true that few unscientific people have this particular type of religious experience. Our poets do not write about it; our artists do not try to portray this remarkable thing. I don't know why. Is nobody inspired by our present picture of the universe?
>
> (Feynman 1964: 5)

But in addition to the scientific adventure, I as a religious scientist feel that I am participating in a human adventure – God has granted me the privilege of participating in the magnificent experiment which he has initiated.

Notes

1 All quotations are taken from the writings of secularists to avoid any possibility of religious bias.
2 A more detailed development in English of Rabbi Dessler's ideas is available in the work of his disciple Rabbi Aryeh Carmell (1975).

References

Alvarez, Luis W. (1987) "Mass Extinctions Caused by Large Bolide Impacts," *Physics Today* July: 24–33.
Aviezer, Nathan (1997) "The Anthropic Principle: What is it and Why is it Important for the Believing Jew?" *B.D.D. A Journal of Torah and Scholarship* 5: 41–54.
Barrow, John and Frank J. Tipler (1986) *The Anthropic Cosmological Principle*, Oxford: Oxford University Press.
Behe, Michael (1996) *Darwin's Black Box*, New York: Simon & Schuster.
Carmell, Aryeh (1975) "Freedom, Providence, and the Scientific Outlook," in *Challenge–Torah Views on Science and its Problems*, edited by Aryeh Carmell and Cyril Domb, Jerusalem: Feldheim, 326–331.
Dessler, E.L. (1954) *Mikhtav Me-Eliyahu*, vol. 1, London: Hoenig (Hebrew).
Dyson, Freeman J. (1970) "The Future of Physics," *Physics Today* Sept.: 23–26.
—— (1979) *Disturbing the Universe*, New York: Harper & Row.
Feynman, Richard (1964) "The Value of Science," in *Science & Ideas*, edited by A.B. Arons and A.M. Bork, New York: Prentice Hall, 3–9.
Gould, Stephen J. (1989) *Wonderful Life*, New York: W.W. Norton.
Hoyle, Fred (1994) *Home is Where the Wind Blows*, Mill Valley, Calif.: University Science Books.
Infeld, Leopold (1980) *Quest*, second edition, London: Chelsea.
Jaki, Stanley L. (1966) *The Relevance of Physics*, Chicago: University of Chicago Press.
Lorenz, E. (1993) *The Essence of Chaos*, Seattle: University of Washington Press.
Popper, Karl (1959) *The Logic of Scientific Discovery*, New York: Basic Books.
Raup, D.M. (1991) *Extinction: Bad Genes or Bad Luck?* Oxford: Oxford University Press.
Simpson, George Gaylord (1967) *The Meaning of Evolution*, revised edition, New Haven, Conn.: Yale University Press.
Spetner, Lee M. (1997) *Not by Chance*, New York: Judaica Press.
Weinberg, Steven (1993) *Dreams of a Final Theory*, London: Vintage.

6   Martinez J. Hewlett

Martinez J. Hewlett is an Associate Professor at the Department of Molecular and Cellular Biology at the University of Arizona in Tucson. He has been awarded research grants from the National Institutes of Health, the National Science Foundation and the Kroc Foundation. Hewlett is a founding member of the St. Albert the Great Forum on Theology and the Sciences. His interests in science and theology led to his first novel, *Divine Blood* (Ballantine, 1998). Hewlett is currently the Southwest Regional Director for the Science and Religion Course Program administered by CTNS.

Interview by Gordy Slack

GS: Would you say a few words about your own religious background?

MH: I was born into a Roman Catholic family and I was educated in Catholic schools through high school. I went to a secular university, The University of Southern California, and after that started drifting away from the faith. It wasn't until quite late in my life – I was in my mid forties – that I came back, in a very dramatic way, to the church. Since that time I have been very active in my faith. I'm very involved in the community at the Newman Center. I'm a lector and a Eucharist minister, and I'm in the process of becoming a lay Dominican, which hopefully, God willing, will occur soon.

GS: How would you describe the transformation that took place in your mid forties?

MH: Epiphanic. I suffered a heart attack and had triple by-pass surgery. In the recovery process I had to sort out what had happened to me and where my life was going at that point. I saw that I had been brought through the trauma of disease and surgery for a purpose. During that period a dear friend of mine – who is also a scientist and a Christian, although not a Catholic – helped me to see what was going on.

GS: The question of purpose comes up a lot in discussions about Darwin and evolution. Some scientists and philosophers insist that religion and religious thinking provide obstacles to thinking clearly about human history and the history of life. In particular, the idea that things happen for some higher purpose. I wonder how you, as a molecular biologist, keep those two ideas going in your life at the same time?

MH: When I came to these questions, I found that I was very ill trained philosophically and theologically. The question of purpose is an interesting one, because most scientists will deny to the death that there is any teleology involved in what they do. But ask any biologist why the chromosomes segregate the way they do during mitosis and meiosis. They will say, "So that the offspring's cells can be the following way." But if you then look at them and say "Well, that is teleological," they will say, "No, of course it isn't. It is strictly mechanistic." I think we are trapped in this underpinning of teleology and trying desperately to be like physicists, which we biologists are not. Some people call it physics envy. I see the purposefulness of creation as an integral part of what I study and what I teach.

GS: When you are looking at a particular scientific problem in biology, you don't try to subtract teleological thinking from your methodology?

MH: I am a molecular virologist. I think in tiny detail about the little intricate parts of viruses. I don't for a minute confuse purpose with that, but I think that it is wrong to feel that there isn't a purpose surrounding all of that and underpinning it. I think science operates on faith without even realizing it. The very nature of my going into the laboratory and doing an experiment demonstrates a faith that there is an essential order which I can observe. If that weren't true, if that faith weren't there, I would have no reason to go into the laboratory.

GS: Is there any link between the points of discovery in your own scientific career and moments of religious insight or religious discovery?

MH: I think most of my scientific career took place during the time between my childhood religious training and this rebirth of religious feeling on my part. When I was working as a reductionist molecular biologist cranking out the data, there were moments of discovery, but I don't think I ever tied them to a religious experience. Since I found my faith again, several interesting things happened, including the fact that I closed my research lab. I wrote my first novel, and I also do theoretical work. I've had a very different experience with science. All of the theoretical things that I've been working on are freshened by this religious insight that is tied to them. It's almost as though I am going back over all of that slowly and fitting it into this larger picture.

GS: In the popular press we so frequently read about a gene being found that suggests an ultimate reductionism – that humans and their behavior boil down to and are determined by their genes. Will this change the way we see human beings and their relationship to the divine and to divine purpose?

MH: Once I was in northern Arizona giving a lecture, making my plea that we are not just simply a collection of genes, and a woman in the audience said her son died of cystic fibrosis. In the hospital, during his final days, a physician said to her, "It's too bad that our science isn't advanced enough that we could take some DNA from your son and create another one for you." She was appalled that he would say that to her. That is where the danger lies. It's mainly the applied molecular biologists who really believe that if I can monkey with your DNA, I can monkey with you.

GS: How do you explain the current fascination with the idea that genes are the sum of what we are about?

MH: I think it is the religion of the day, this faith that molecular biology will solve all our problems. It touches something primal and essential in us, but it's the high tech flavor that everybody likes, with buzzing and flashing lights and everything. We have this dream of "Can we do that, can we play God?" I think that's the fascination in the popular mind – that scientists are in the laboratory and they are actually going to design a better me.

GS: The idea that we could finally know ourselves in some fundamental way is also very attractive.

MH: I'm not sure if, in the popular mind, it's really a desire to know themselves so much as it is a desire to be able to control what happens. Somebody who is dying of cancer wants very much for you to go in and turn that gene off. Give me an injection of this virus if I've got a tumor and it will just destroy the tumor and I'll be free. That's the kind of thing people are drawn to: the promise that what scares them most – dying – can be staved off.

GS: Do you think that neurobiologists' work on the brain is narrowing in on the self?

MH: Well, so long as the area of study is the physical, by definition, all you will ever see is the physical. How could you see anything that I tell you is spiritual in aspect? It will always be thus, until somebody is willing to admit that the spiritual is a different realm. This is what is called scientism: the belief that science not only is powerful but also defines all knowledge. Anything that doesn't get defined by science is therefore not knowledge or is not real. Scientists are seduced into thinking that our science has the complete view. The only reason we think it has the complete view is because we don't know any others.

GS: Do you think that the seductions of science and of material reductionism have permanently siphoned energy and attention away from religion and spiritual life?

MH: My colleagues, those in my age group and perhaps a little bit younger than I, are seduced. But the students that we are talking to now are much more open to other possibilities. That openness can lead them in a lot of very strange directions, questioning of everything, without any kind of foundation. But I see in them a willingness at least to accept that there may be something more than they are being shown in their textbooks. I was in a lecture one day and one of my colleagues happened to be visiting. A student asked me why a cell did something. It was one of those places in science where we don't really know, so I did a God of the Gaps answer and said "Because God made it happen." Everybody laughed. When we were leaving the lecture hall, my colleague pulled me aside and said "You know, you shouldn't have said that."

I said "Well, what should I have said?"

He said "You should have said Mother Nature made it happen that way."

I have a sense from this generation of students that there is a chance they may view things differently than we did. That means there may be permission for them to be scientists and be spiritual at the same time.

Martinez J. Hewlett

What price reductionism?

Introduction

At the start of the twenty-first century it is probably correct to say that we are at the pinnacle of the scientific program envisioned by the philosophers of the post-Enlightenment. When David Hume proposed that the material world be the strict domain of the scientist and that empiricism should be the approach to understanding that world, an experimental method that had begun with Galileo and had been refined by Newton and Bacon was embraced with fervor. The Cartesian dualism that inspired this separation of the material and non-material aspects of reality into different spheres of human inquiry clearly led, in a pragmatic sense, to the progress of modern experimental science. But how did the exclusion of the non-material or spiritual aspect from consideration by science lead to the denial of its existence that has become a hallmark of the current view? Is this denial a reasonable result of the scientific enterprise? I will argue that the philosophical stance of science in which the spiritual not only is excluded from investigation but is *de facto* deemed to be impossible stems from two features of the research program followed in the natural sciences: reductionism, and the replacement of classical causality with scientific causality. For purposes of this argument, examples will be taken mainly from the biological sciences.

A molecular biologist looks at modern biology

How is it that someone trained in modern molecular biology, who is also a virologist, concerned with the smallest of things that exist at the border between the living and the non-living world, would come to question the central place held by the reductionist approach?

I have been associated with the St. Albert the Great Forum on Theology and the Sciences, currently entering its eighth year at the Catholic Newman Center of the University of Arizona. I was asked to be on the advisory board of the Forum and have served there since its founding. Originally, I had little interest in philosophy, seeing my purpose in planning this series as the "voice of reason and science."

The Forum has invited distinguished scientists, philosophers and theologians to present their ideas in an atmosphere of Christian discussion. One visitor early in the series was Fr. Benedict Ashley, the great Dominican philosopher and moral theologian. During his presentation he spoke of the nature of living things, emphasizing their wholeness. He used a cat as his example.[1]

At one point I rose to object to Fr. Ashley's characterization of living systems.

"I could take an organ from the cat, extract the DNA from the cells, sequence that DNA and, after computer analysis, I could tell you that the DNA is 'cat'."

I, of course, spoke with all the arrogance of my profession.

Fr. Ashley, however, replied immediately.

"It's no longer a cat."

I remained self-righteous.

"When did it stop being a cat?"

"When the cat died," he replied.

His answer left me stunned and propelled me on the journey into the philosophical underpinnings of modern science upon which I find myself today.

Reductionism in modern biology

The evidence suggests that molecular biology has finally "made it" on the popular scene. DNA took center stage in Michael Crichton's novel and movie, *Jurassic Park*. More recently, the double helix, or a fragment thereof, has main title billing in *GATTACA*.

The story presented here describes a future world in which the entire progress of a person is dictated by the knowledge of his or her genetic information. With the sequence of As, Gs, Ts, and Cs known at birth, all of the possibilities for the life of the individual are "predicted."

GATTACA represents the ultimate nightmare for a society that fears modern biology as reductionist. And yet this very approach has been extremely useful and productive for science. Why should this raise the specter of a world such as described in this movie?

Let us first define our terms. The process of reduction consists in explaining a phenomenon studied at one level of reality by the principles and processes understood at the level of the component parts of the phenomenon (Agazzi 1991). For example, the explanation of all of chemistry in terms of atomic physics, or the explanation of all biological phenomena in terms of chemistry and physics.

In a very practical sense, the empirical approach that has characterized science since Galileo and, more importantly, since the Cartesians would not have been possible without the use of reduction. In a formal sense, methodological reductionism attempts to frame all sciences in the guise of mathematical physics, with quantitation, deduction from general laws, and what may be called verifiable – or at least falsifiable, in the view of Karl Popper – predictions (Agazzi 1991). In practice, it is simpler experimentally to observe one or two of the components of a thing in a reproducible fashion than to observe the entire thing in action. Further, the level with which the discipline is concerned, especially in the mixed sciences, may dictate a methodological consideration of the components. For instance, the proper province of biochemistry is the study of enzymes within a cell.

The tremendous power and success of modern science owes much to this reductive approach for science's understanding of the physical world. And yet it is the very seductive nature of this power that results in serious problems when misapplied. It is quite easy for the scientist to conclude that the only valid way of gaining knowledge of the physical world is by reducing a complex phenomenon to the structures and principles that describe the components of that phenomenon. This conviction is called epistemological reductionism (Agazzi 1991). When knowledge of the physical world gained in this way leads a scientist to conclude that everything about reality can be completely described in terms of the most basic components of that reality, a metaphysical statement is being made by that scientist. This is called ontological reductionism (Agazzi 1991).

Within biology these three forms of reductionism (methodological, epistemological, and ontological) have varying utility. Clearly, methodological reductionism has been essential for the progress of fields such as biochemistry and molecular biology. The very nature of these "mixed" disciplines requires that aspects of living systems be examined at the level of their chemical and molecular components. Add to this the quantitative and deductive mind-set captured from physics and the picture is complete. From reductionism has come the great scope of modern biology, with its descriptive and explanatory approach that begins with the chemical nature of the gene and includes the view of the cell under the governance of the information residing in the sequence of nitrogenous bases that form the DNA.

As the progress of modern biology continues, it may be easy for the scientist involved in the enterprise to look for more detailed information using the same approach. In fact, it can become the sole approach upon which experimental designs are based. Thus, the great enterprise of the Human Genome Project has, as its premise, the epistemic conviction that the determination of the genomic location and base sequence of the set of all human genes is prerequisite to understanding their functions. Certainly, for the goal of gaining knowledge about the specific molecular subunits that constitute a living system, this research program will be successful. But can this project completely describe what it means to be human?

For some biologists, the very nature of living things must be reduced to the properties of the fundamental particles of which they are made. Francis Crick, the Nobel laureate who, along with James Watson, proposed the currently accepted structure for DNA, wrote: "The ultimate aim of the modern movement in biology is in fact to explain all biology in terms of physics and chemistry" (Crick 1966: 10).

While this may, at first glance, seem to be a logical extension of the methodologic and epistemic programs set out above, this statement in reality constitutes a philosophical position.

If the plan put forth by Crick were to examine the structure of DNA and to attempt to explain the physical properties of this molecule in terms of quantum-mechanical principles, then there would be no difficulty. This effort would merely be a methodologic decision to attempt a quantitative description of the collection of atoms that constitute the DNA molecule, based upon, for instance, approximations to the wave equation for the individual components. Such quantum-biochemical descriptions are not philosophical statements but refinements of the chemical description of this material.

If, on the other hand, the statement purports that all aspects of living systems are explainable by these quantum-mechanical descriptions, the intent is quite different. At the heart of this approach is the belief that nothing exists other than the physical aspects of reality and the correlate belief that all properties of that physical reality result completely from the properties of the simplest elements of that reality. This is clearly Crick's intent, as he goes on to say: "Eventually one may hope to have the whole of biology 'explained' in terms of the level below it, and so on right down to the atomic level" (Crick 1966: 14).

This position is a metaphysical one, in that it proposes a statement about the nature of reality. This may easily be described as ontological reductionism.

Causality and modern science

The classical and modern scientific ideas of causality are quite distinct. In the classical sense, causality indicates dependency. A particular effect is dependent upon a particular cause. Modern science uses the term causality to imply predictability. A particular effect is predicted from a given cause.

In the classic view there are four kinds of causes: the material cause or matter (here "matter" is not used in the same sense as it is in physics or chemistry), the formal cause or form, the efficient cause or agent, and the final cause or end. These causes and descriptive biological examples are outlined in Table 6.1 (with definitions adapted from Wallace 1996 and Dodds 1995). This classic notion of cause and effect was uprooted through the skepticism of Hume, who held that we can have no knowledge of this kind of relationship. His view, which has influenced the philosophy of our scientific enterprise, was that events ("effects") can only be sensed or experienced sequentially and that no relationship between them can be known. Thus, we may often experience one event (an effect) following another event (a cause) but we cannot know that the cause will always lead to the effect. In this case, the idea of an efficient cause (agent) and final cause (end) must be discarded. Hume used the term "causation" to describe this sense-derived expectation of a sequence of events.

If this is the case, how did modern science stray away from this Humean ideal and come to the concept of predictability? In a strict sense, "predictability" implies "determination." Determinism flows from the Newtonian view of the physical world that has only been upset this century by the rise of quantum theory and relativity. In spite of the changes that have occurred in physics, biology has in many ways clung to this deterministic view. Thus it is that "causality" is seen as the ability to predict an effect from a cause.

Another inheritance from the philosophy of Hume is the restriction that the only valid knowledge of reality comes through sense experience during an examination of the material aspects of the world. Hume's rejection of metaphysics included a rejection of possible consideration of any spiritual dimension of reality.

In the modern sense, then, the aspects of the proteins α and β tubulin can be investigated with respect to the material out of which they are made (ultimately, the material cause) as well as the structure which these proteins take (in effect, the formal cause). In no case, however, does modern biology speak of the efficient cause in the classical sense or of the final cause (teleology) of the microtubule subunits. The nature of the structure of the tubulin subunits that promotes their assembly into microtubules is viewed as the "cause" of the formation of the spindle fibers, but not in the classical meaning of the term. In addition, if asked why the spindle fibers form, a biologist might answer "so that the chromosomes will be correctly distributed into the daughter cells." And yet, if told he or she is making a teleological statement, this will be vehemently denied.

Table 6.1

Cause	Definition	Example
material	that which causes a thing to be; a potential or possibility; proto-matter	the material essence of things, for instance, of proteins
formal	that which causes a thing to be a certain kind of thing	a particular protein, for instance, α or β tubulin
efficient	that which causes the relative state of motion of a thing; the agent of change	the nature of α and β tubulin that allows assembly into microtubules
final	that which is the purpose or sake for which something is done	microtubules organize into spindle fibers that result in correct chromosome movement during cell division

Steps to the scientific rejection of the spiritual

Much of modern biology and perhaps of modern science in general rejects the notion of the spiritual as a possible aspect of reality. For the case of biology, the scientific philosophy that allows this includes the following features: a limitation to exclusive consideration of the physical or material; a rejection of efficient and final causality in the classical sense, in spite of obvious teleological implications of observations; a deterministic causality coupled with a Humean notion of causation; an appeal to physics as the ultimate basis, by reduction, of biological phenomena.

As a case in point, let us examine the developments that have led to what has come to be called neo-Darwinism and the conclusions concerning God which some have erroneously drawn from this molecular view of the evolutionary process. Table 6.2 describes the sequence of events that have occurred since Darwin first published his book *On the Origin of Species* in 1859.

Notice that each of the events described in the table, up to the last two, are well within the province of scientific investigation. The conclusions drawn in these cases derive from experimental observations, hypothesis-formation and testing, as well as theory-construction. In addition, the progress of the enterprise towards the identification of the gene as the physical thing upon which the force of natural selection acts is one characterized by a

Table 6.2

Event	Method
Darwin's observations lead to the hypothesis of natural selection as the operative force in evolution (1859)	observation → hypothesis/theory
Mendel quantifies inheritance and proposes the gene as the unit of inheritance (1868)	observation → hypothesis/theory
DNA is shown to be the genetic material (1944–1954)	observation → hypothesis/theory
Genetic changes are the result of mutations	observation
Mutations are random events that result in changes in DNA sequence	observation
Mutations are proposed as the way in which variation occurs, upon which variations natural selection acts	induction
It is not necessary to postulate an intelligent creator to explain the results of random events	metascience
Science (through Darwinian evolution) proves that there is no God	metascience/philosophy

reductive view of biological systems. These events are then coupled with the prevailing philosophical position underlying modern science that the spiritual not only is excluded from scientific investigation but also does not exist. The result is a mistaken view that since mutations are "chance" occurrences and therefore unpredictable (notice the notion of causality in force here) no intelligence is required for their occurrence. Given the *a priori* stand that God does not exist, the evidence of biology is taken to prove the circular argument that God, in fact, does not exist.

The nature of this metaphysical conclusion is exemplified by Daniel Dennett:

> A familiar diagnosis of the danger of Darwin's idea is that it pulls the rug out from under the best argument for the existence of God that any theologian or philosopher has ever devised: the Argument from Design. What else could account for the fantastic and ingenious design to be found in nature? It must be the work of a supremely intelligent God. Like most arguments that depend on a rhetorical question, this isn't rock-solid, by any stretch of the imagination, but it was remarkably persuasive until Darwin proposed a modest answer to the rhetorical question: natural selection. Religion has never been the same since. At least in the eyes of academics, science has won and religion lost. Darwin's idea has banished the Book of Genesis to the limbo of quaint mythology.
>
> (quoted in Brockman 1995: 187)

It must be emphasized that the error is not in the science which, given the current level of understanding and available tools, is based upon relatively sound logical foundations. Rather, the problem here is a philosophical one, wherein the scientist has overstepped the agreed upon limitations of the endeavor and entered into a discussion for which he or she has no logical basis. The fact of the existence of a spiritual dimension to reality or of an intelligent Creator is certainly outside the sphere of examination by science *per se*. And yet the full weight of the scientific method is often brought to bear against these ideas, as in the quote above from Dennett.

Conclusion

The methodology of modern biology is generally reductionistic and deterministic, in spite of the changes which have taken place within physics, the science to which some strive to reduce biology. This "bottom up" causality can lead to a number of conceptual difficulties in light of the observed properties of living systems when compared to their isolated components. Examples abound in the sciences of the inability of the tool of reductionism to answer a particular question. For instance, the chemical properties of the sodium atom cannot be predicted from the properties of an isolated electron. Similarly, the development of multicellular creatures such as human beings is not simply the result of the molecular program found within the sequence of bases in DNA. The physical development itself, to say nothing of the spiritual dimension, is much more complex and requires an inter-acting system of cells to achieve the ultimate form.

The anti-metaphysical stance of the scientific enterprise, inherited from the eighteenth-century philosophers, permits no consideration of the spiri-tual as a part of reality. For these reasons some biologists mistakenly conclude, by circular reasoning, that the accumulated evidence disproves the existence of the spiritual and of God. It is especially important when biolo-gists take this stand since, unlike cosmologists or physicists, they are looking at human beings as organisms within creation. I contend that there is no necessity for this extreme position and, moreover, that it harms the scientific enterprise by greatly limiting the possibilities. Investigations of material reality should be informed by, not at odds with, investigations of spiritual reality, and vice versa. There is no need for scientists to overlook what is perfectly obvious in everyday experience, even if it cannot be measured by instruments.

Note

1   It is interesting to see the role played by cats in philosophical discussions, as witnessed by Schrödinger's famous and ill-fated (one-half of the time) cat.

References

Agazzi, Evandro (1991) "Reductionism as Negation of the Scientific Spirit," in *The Problem of Reductionism in Science*, edited by Evandro Agazzi, The Netherlands: Kluwer Academic Publishers, 1–29.

Brockman, John (1995) *The Third Culture*, New York: Simon & Schuster.

Crick, Francis (1966) *Of Molecules and Men*, Seattle: University of Washington Press.

Dodds, Michael (1995) Course notes for "Philosophy of Nature, PH1056," Berkeley, Calif.: Graduate Theological Union.

Wallace, William (1996) *The Modeling of Nature: Philosophy of Science and Philos-ophy of Nature in Synthesis*, Washington, D.C.: Catholic University of America Press.

7   Robert B. Griffiths

Robert B. Griffiths is Otto Stern University Professor of Physics at Carnegie-Mellon University, where he has been a faculty member since 1964. He has a Ph.D. in Physics from Stanford University (1962). He is a Fellow of the American Physical Society and a Member of the National Academy of Sciences of the USA. His current research interests include the foundations of quantum mechanics, quantum information theory, and quantum computation. Earlier he worked on the mathematical foundations of statistical mechanics, the theory of critical phenomena, models of magnetic materials, and the theory of incommensurate crystals. His other interests include: the problem of determinism and free will, and the relationship of physical science to Christian theology.

Interview by Philip Clayton

PC: Let me ask you first about your religious background and your own religious commitment.

RG: My parents were Presbyterian missionaries in India, and I'm currently a member of a Presbyterian church, although I have probably moved somewhat toward Baptist theology.

PC: Was there any influence of the pluralistic religious context of India in your own formation as a Christian?

RG: I would find it hard to say that there was. I didn't start seriously thinking about my faith from an intellectual point of view until I got to college in the USA, and at that stage Hindu ideas were far removed.

PC: Do you see your faith and your scientific practices as relevant to each other?

RG: I think people bring to their religious concerns many different styles or modes of operation. What I tend to bring is my training as a scientist, and thus the first thing that's important is: is something true, and what sort of system does one have for thinking about it? I sometimes consciously, but more often not so consciously, bring a lot of my prejudices as a scientist into my religious practice. A certain frankness and openness is, I think, a scientific trait. You tell people why you disagree with them, and that you think they're wrong, and often do this with a lack of sensitivity. I think that's both good and bad in religious practice. But generally, I think bringing to the religious community some of the positive aspects of the scientific community is a good thing.

PC: What areas in Christian practice do you think are radically different from scientific practice?

RG: Faith is one difference, but not totally different, because modern scientists believe in quarks, which are in principle invisible. But religion, or at least Christian theism, says that a Christian's obligation is to serve and worship God. God is the moral authority, and the physical universe around us has no moral authority.

PC: Is faith something that means accepting on the basis of authority a body of truth, and therefore is highly disanalogous to science, which accepts very little on unquestioned authority?

RG: Scientists accept a great number of things on authority. The freshmen that I'm currently teaching believe certain things because the book says so, and I say so, etc. One wants to balance off their acceptance of authority, without which they'll never learn the subject, with an appropriate amount of intelligent fooling around and asking questions, including dumb questions. So both authority and doubt play a role in physics.

Now what do we do as Christians? Certainly a great deal is received on authority from parents and pastors, who pass along the teaching of the Bible, and what the community has produced through the years. But one is free to challenge the tradition handed down, go and think about it, and argue about it. Certainly, one can go off the deep end in the same way that freshmen sometimes come up with stupidities, but I myself feel that the questioning approach is quite appropriate. When I read the Psalms I find honest questioning. The psalmist says, "My God, my God, why have you forsaken me?" It seems to me there is some parallel between that and scientific questioning.

Now, I believe that God's authority in the moral realm is a very different thing from what one encounters in science; there's nothing comparable to that in physics. I don't see a problem with this, because to me the very nature of God means that you have to relate to him in a very different way than you relate to the physical world. I accept the Bible because I believe that God spoke in particular ways to the people whose writing one finds there. And there's a great deal about personal relationships, about how you should behave toward people, to be found in the Bible, and you can go and test it out, see whether it works.

I think that there is a reality "out there." God and moral law and so on are real things which exist. For me, religion should be based on a reality which is "out there," whose truth does not depend upon whether or not we like it or believe it.

PC: For you, are there any developments in the physical or biological sciences which might falsify Christian belief or even count against it?

RG: My way of thinking about Christian belief is to regard it as something which can be discussed, argued about and (potentially) falsified. I don't think that Christianity stands off in some domain by itself, and that it could not be falsified. But it's in the historical sciences, not the physical, that I can at least imagine developments which could falsify Christian belief.

PC: In classical Christianity, like in Judaism and Islam, there was a stress on God's presence in the universe and activity in the world. God was understood to bring about changes in the physical world. Do you see this notion of God's activity in the world as being unaffected by the growth in the physical sciences, or being challenged by it in any way?

RG: I think that God is at work in the world in terms of normal events, because if you believe that he creates the world in space and time, then in some sense he is the origin of all that happens. I say "in some sense," because this belief leads, as is well known, to all sorts of intellectual problems and troubles: fights over determinism and free will, the problem of evil, and so on. Then there are the far-from-normal events which are ascribed to special divine activity. Let's take the case of Jesus walking on the water as reported in the Gospels. Walking unsupported on the sea of Galilee violates the laws of hydrodynamics and therefore it's an impossibility. But the Christian can reply that, after all, Jesus was the incarnate Son of God, the Creator come to live among us.

Given that extraordinary state of affairs, should we not anticipate a few things which don't follow the regular pattern of God's activity as we see it in natural laws? I think that all that contemporary science is able to do is to point out that certain wondrous acts are contrary to the normal behavior of the world. Certainly walking on water violates the laws of hydrodynamics, but hydrodynamics isn't a fundamental law of nature. So far as the laws of quantum mechanics are concerned, the probability of walking unsupported on the surface of water is extremely small, but not impossible. Is God really working in the world? If he is, then, almost by definition, he can do that which is improbable. He controls the probabilities. We roll the dice, and God decides how they fall.

PC: You specify a way of thinking about divine action that would be physically consistent, namely quantum intervention by God. Yet you speak of a habit of mind which is disinclined to accept claims about miracles. Is that sort of tension widespread, not only among scientists, but among Christian believers in the United States?

RG: It seems likely that there has always been a certain amount of tension when people have reported the occurrence of miracles – when Paul was preaching in Athens, were his listeners any more ready than we, who live in the twentieth century, to believe that Jesus rose from the dead? I doubt whether they were.

PC: I want to end by asking a slightly more personal question, namely whether you've derived any religious inspiration from your scientific work?

RG: I would certainly affirm that science provides grounds for worshiping the Creator. I think that it's just glorious what modern science has shown us about the world. If the Hebrew poets could praise God for the wonders of his creation, I think I ought to be able to do it even better – though my poetry is far beneath their standards.

PC: So you say that knowledge of the physical world and the beauty that we discover in the mathematics and in the actual phenomena themselves is, for the Christian, inspiring?

RG: For myself as a Christian, yes. There are Christians who say the world is only 10,000 years old, or that evolution cannot possibly have occurred. Aren't they going to be ashamed, come the day of judgment, to have to say to God, "We saw your wonderful works, but we decided we couldn't believe in them!"

PC: Are there any ways in which your faith as a Christian theist has informed or motivated your work during your career?

RG: Physical science deals with physical objects in the world, and constructs theories about them, but it is carried on by a community of people, and thus all the issues of relationships, of integrity and honesty and personal goals come up and have to be dealt with. I personally have found it very beneficial to be able to bring to my own participation in the scientific community a knowledge of the Bible, and beliefs and convictions about what is right and wrong, and how to treat people, which have a Christian foundation. I don't think I've carried those convictions as far as I should have in terms of compassion and concern about other people. I have sometimes been more ready to argue for theological doctrine than to actually give someone a helping hand. I am, nonetheless, thankful for my Christian upbringing, for support by the Christian community, for instruction in the Bible, and for the people who pray for me and with me, because being a scientist does not mean one ceases to be a human being!

Robert B. Griffiths

Scientific reduction

Adversary or ally?

Introduction

It is widely believed in debates on science and religion that scientific reduction, meaning the explanation of complex phenomena in terms of simpler components, is a threat to religious belief and a support for the materialist view of the world which claims that matter is all there is, and that nothing else is needed in order to understand the universe. The purpose of this essay is to call that belief into question.

Reduction, in the sense of explaining complicated phenomena in terms of simpler components, is very much a part of modern science. It characterizes a great deal of what has gone on in twentieth-century physics and chemistry and is playing a more and more important role in biology. In particular, we do not nowadays ascribe the properties of living organisms to some sort of special "vital force" which is absent from the realm of non-living phenomena. Instead, we think of biological organisms as made up of cells, and cells as complicated structures in which chemical reactions and physical processes obey the same laws which chemical engineers use to design objects having nothing (directly) to do with living systems. Chemical reactions are, in turn, regarded as manifestations of the fundamental principles of quantum mechanics when these are applied to systems of electrons and nuclei interacting through electric and magnetic forces. Granted, the program of reduction is not complete, since the way in which the properties of electrons, quarks, and the like arise out of some more fundamental structures, assuming these exist, is not yet understood, and gravity cannot yet be comfortably accommodated in the quantum world. Nonetheless, one has to concede that there has been an enormous amount of progress in the last hundred years; recall that at the close of the nineteenth century there was still serious doubt about the very existence of atoms, and essentially complete ignorance of their internal structure.

But are there any limits to this program of scientific reduction? If human bodies and human brains are composed of cells, and their operation can be described in terms of physical and chemical processes, is there any room for the human soul? What about the doctrines of immortality and the resurrection of the body? Even those who would gladly dispense with these in the name of clearing away superstitions left over from a pre-scientific age may be troubled by the question of how chemical machines can make those free choices which seem to underlie our notions of human agency and moral responsibility.

For traditional religious belief there is the additional issue of how, if the entire universe is simply a set of atomic particles governed by deterministic laws, there is room for God to act in response to prayer, or in order to perform miracles. To be sure, modern quantum theory has gone a long way in undermining classical ideas of determinism, but replacing determinism with random processes raises questions about whether, and if so how, God governs the world, and whether there is any basis for the hope for the future found in traditional theism.

One way to counter the threat of scientific reduction is to argue that it does not and will not work: there are limits on what can be achieved, and eventually the program of reducing the human body to physics and chemistry, or human thinking to interactions among neurons, or the human race to a product of evolutionary development, will fail because that is not the way the world really is, or was. It is true that scientific studies have not demonstrated beyond any reasonable doubt that humans evolved from lower animals, or that the human brain operates according to the usual laws of biochemistry, and claims to the contrary are premature. It is entirely proper for critics to raise the issue of how far a given program of scientific reduction has actually succeeded. A disinterested examination of the evidence will generally reveal that progress toward a hoped for "final theory" is a lot less than the enthusiasts claim, though it is often more substantial than the critics will allow. Discussion, argumentation, criticism, and response are all part of the scientific enterprise. Scientists and their critics can both benefit from greater humility.

However, in this essay I want to address a somewhat different question. Suppose some program of scientific reduction were to be extremely successful, what would be the result? For example, if at some future date it should prove possible to relate human consciousness and thinking in a very precise way to the operation of neurons in the brain, in the same way in which we at present understand the operation of computers in terms of transistors and electrical circuits, what would this imply? Would human freedom and moral choice simply disappear, suffering the same fate as phlogiston and the geocentric universe?

From the way I have raised this question, it will be apparent that my own answer is "No."[1] My approach in this essay will be first to discuss what I regard as a very successful case of scientific reduction, that of thermodynamics to atomic physics. Then I will explore whether lessons learned from studying this example can provide some perspective on a different program: scientific reduction of human thought to biochemical processes in the brain. One advantage of starting with thermodynamics is that, so far as I know, there are no important religious or philosophical issues tied to the outcome, so one can discuss it without becoming emotionally involved. Also, since the main developments lie in the past (or so it seems), one does not have to guess what they are going to be.

❡ The reduction of thermodynamics to atomic physics

The science of thermodynamics was developed during the nineteenth century at a time when the existence of atoms was very much in doubt, since all the evidence for atoms and molecules was quite indirect. Nonetheless, efforts were made to explain the first and second laws of thermodynamics in terms of the mechanical properties of atoms and their interactions. These made use of classical mechanics, since quantum mechanics had not yet been developed. The development of quantum theory in the 1920s meant that earlier discussions had to be reopened. There is a continuing debate and discussion about these matters, both in classical and quantum terms, although it is hardly a topic in the forefront of physics research nowadays.[2]

The first law of thermodynamics states that energy is conserved, and for this law the reduction to atomic physics has been quite successful. The basic idea is quite simple: given a mechanical system of interacting atoms (or molecules, or whatever; for simplicity I will speak of atoms), it is possible in classical mechanics to ascribe a kinetic energy to each atom. The total mechanical energy is the sum of the kinetic energies of all the atoms, plus the potential energy associated with their interactions. This total energy can be identified with the thermodynamic energy of the system composed of these atoms, and the first law of thermodynamics is then an immediate consequence of the conservation of potential plus kinetic energy in mechanics. One can, to be sure, still ask why mechanical energy is conserved in the case of atoms. At present, energy conservation is simply a fundamental postulate, one which has been amply and abundantly verified by all sorts of experiments, but which cannot be derived from more fundamental principles. One can certainly imagine universes in which energy is not conserved, but we do not seem to live in one. Anyway, admitting that there are still unanswered questions in the science of mechanics does not indicate any defect in the program of reduction as such. We can still say that we understand the thermodynamic law, and indeed can derive it, from a basic law of mechanics.

This discussion must be modified in certain details, but remains fundamentally correct, if the mechanics governing atomic motion is quantum mechanics. Once again, it is possible to identify a total energy of the system, which in quantum theory is associated with the Hamiltonian operator. Splitting it into kinetic and potential energy contributions turns out to involve some subtleties, but in any case that division is not needed for thermodynamics, since the thermodynamic energy can again be identified with the total quantum energy. Conservation principles and even the ability to state that a system has a definite energy are subtle points in quantum theory. But because thermodynamics provides a rather coarse description (by the standards of atomic physics) of large systems, such subtleties do not pose serious problems for the program of reduction. Or, to put it another way, the problems which still remain in explaining the first law of thermodynamics in terms of atomic physics are part of the general, and somewhat controversial, problem of providing a satisfactory physical interpretation of quantum theory, and thus do not reveal any obvious defect in the program of reduction as such.

The second law of thermodynamics states that the entropy of an isolated system increases with time or, once the system has reached equilibrium, remains constant. Thus it singles out a particular direction of time, and processes governed by the second law are said to be (thermodynamically) irreversible: they proceed in one direction, but not the reverse. Attempts to understand the second law in terms of atomic processes began with the work of Maxwell and Boltzmann during the last century. While much progress has been made since then, a number of problems remain which have not been satisfactorily solved, at least by the standards of the (relatively small) community of scholars interested in such things. Examining both the more successful and the more problematical parts of this program of reduction will help to provide some useful insights into what might be expected in other cases of scientific reduction.

Entropy, unlike energy, has no direct mechanical counterpart, and in a sense this is at the heart of the difficulties which have been encountered in providing a mechanical explanation, based in atomic physics, of the second law. During the nineteenth century, the science of statistical mechanics was developed by (among others) James Clerk Maxwell, Ludwig Boltzmann and Josiah Gibbs, and it is by the use of statistical mechanics that we think we currently understand the second law in atomic terms. Statistical mechanics means applying laws of probability theory to mechanical systems, and that this appears in the reduction process is itself of some interest. For it was not at all evident at the outset that introducing probabilistic ideas would be a useful route to follow in trying to understand the second law. Remember that, unlike the situation in quantum theory, there is nothing inherently probabilistic in classical mechanics. Thus it is not surprising that Boltzmann's attempts to explain the second law in probabilistic terms were initially rather unclear, and met with heavy criticism. Nowadays, the use of probabilistic ideas in classical mechanics is more familiar, since they play an important role in the study of chaos. But even so, there are still many loose ends which remain in the effort to understand the second law in probabilistic terms, whether one uses quantum or classical statistical mechanics (see Sklar 1993).

One of those loose ends is irreversibility: why does entropy always increase in time? There are informal arguments which, at least in my opinion, contain the essential ideas, but these need to be cleaned up. As Sklar (1993) makes clear, mathematical physics has not yet provided results

for non-equilibrium statistical mechanics of the same character and generality available in the equilibrium case. We are quite confident we know how
to calculate the entropy for a system in equilibrium, but the precise procedure for a system out of equilibrium is not clear, and we don't have a
satisfactory demonstration that this entropy will, with high probability,
increase with time.

Let us assume for the present discussion that tying various loose ends will
not produce any radical changes in our present understanding of how thermodynamics is related to atomic physics. Then one can say that the two laws
of thermodynamics are understood in quite different ways from an atomic
perspective. The first law follows from a relatively straightforward application of a conservation principle which plays a fundamental role in both
classical and quantum mechanics. One can speak about the energy of a
single particle, of two particles, of ten particles, or of ten to the twentieth
particles, and there is no conceptual gap on the way from the atomic physics
of single atoms to the thermodynamic law.

The second law is very different. It doesn't make much sense to talk about
the entropy of a single particle; while one can write down a formula for it
under some circumstances, it isn't clear that it has much significance.
Instead, entropy has something to do with the behavior of large numbers of
particles, under circumstances which we do not as yet fully understand, but
which seem to be fulfilled in many different objects we encounter in our
everyday experience, as well as in the laboratory. Entropy is, one could say,
an "emergent" property which does not apply to a small number of particles, but is extremely useful for describing large systems of particles. Perhaps
the concept of a "government" provides a useful analogy: the term does not
make much sense for a single, isolated, human being, but is very useful when
one thinks about societies containing substantial numbers of people.

Furthermore, in the process of reducing the second law to atomic physics,
it has been necessary to invent a new discipline, statistical mechanics, or to
put it another way, it has been necessary to introduce new, probabilistic
concepts into the discipline of mechanics, concepts which are not, at least in
classical mechanics, an intrinsic part of the subject. These new concepts
appear to be essential to the process of reduction. For example, both classical and quantum mechanics do not single out a direction of time,[3] whereas
the second law of thermodynamics does single out a direction of time, the
direction in which entropy is increasing. The only way I know of for reconciling these seemingly contradictory perspectives involves, among other
things, invoking probabilistic descriptions: thermodynamic irreversibility is
not guaranteed by the laws of mechanics, but it is rather likely.

But should we think of entropy as part of the real world? Could it not be
just a useful fiction employed by scientists and engineers as part of their
calculations? Is entropy as real as some material object, say a brick, or the
atoms of which the brick is composed? Even if there are problems reducing
entropy to atomic physics, is this anything a good materialist should worry

about? Rather than attempting to directly address a question which is full of philosophical difficulties, I shall adopt the coward's tactic of raising an alternative question of the same sort.

The typical man on the street, at least if the street is in Pittsburgh, has never heard of entropy, but he knows we have an "energy crisis," and that our weather is uncomfortably cold in the winter. For the thermodynamicist, entropy, energy and temperature are all closely related.[4] As a consequence, whatever difficulties arise in reducing entropy to atomic physics are also present in the case of temperature. This last point is easy to overlook, because in classical statistical mechanics temperature is related to kinetic energy in a simple way, so that relating thermodynamic temperature to atomic physics seems no more difficult than the corresponding problem for the total energy. However, this is one instance in which quantum theory really does make a difference: the relationship of the kinetic energy of a quantum system to the temperature is not at all simple, and thus one is back to using entropy, or something equivalent. Consequently, to the person who has doubts as to whether entropy is part of reality, I confess to worries of the same sort, but I add that my worries also apply to temperature. And whatever subtleties may be involved in understanding the latter in terms of atomic physics, as a practical matter I find it prudent to have a warm coat, hat and gloves available when wintering in Pennsylvania!

Has thermodynamics been "reduced to" atomic physics? That depends upon what one means by "reduced." Even taking an optimistic view of what has been accomplished to date, with the expectation that the various loose ends will eventually be tied up, or could be if someone were sufficiently interested, there is no reason to suppose that the discipline of thermodynamics could or should be replaced by atomic physics. Thermodynamic laws are extremely good descriptions of what is going on in the world. They are "phenomenological" in the sense that they refer to everyday things accessible to macroscopic human observation, and make no reference to what is going on at the atomic or molecular level. Just because we have atomic explanations for thermodynamic laws (remember, I am taking the optimistic perspective) does not mean that those laws are wrong, or that the entities they refer to are unreal, either in practice or "in principle."

Furthermore, while one cannot claim that thermodynamics or other phenomenological theories, such as hydrodynamics, are essential if we are to understand the natural world – they might someday be replaced by something better – there are excellent reasons to think that they *cannot* be replaced by some sort of detailed atomic description. Naively, one might suppose that all we *really* need to know are the positions and velocities of the atoms, along with the laws governing their interactions, in order to have a "complete" description of nature which would not need the concept of energy or entropy. But such a precise mechanical description, for even a modest number of atoms or molecules, say those in a cubic centimeter of air, requires such an enormous amount of information that there is no way in which it could be stored on any present, or any conceivable future computer. To be sure, nothing prevents us from imagining, as did Laplace, a super-intelligence who in a manner inconceivable to us somehow accumulates the requisite information and thereby manages to understand the situation. But since that understanding evidently cannot be communicated to us, it seems somewhat misleading to claim that a thermodynamic or hydrodynamic description can "in principle" be replaced by a detailed atomic description. Things do not improve if we suppose the super-intelligence uses quantum mechanics rather than Laplace's classical mechanics.

In addition, we are not able to derive the second law from atomic physics without invoking additional, probabilistic ideas. To the non-scientist, this might look like a serious flaw in the program of reduction. The research scientist is likely to have a different attitude, because he or she takes delight in introducing some new idea in order to explain what was not previously understood, or at least not understood as well as is possible using the new idea. The program of reduction, viewed in this way, is not a sledgehammer to demolish prior knowledge, an adversary of whatever was known in the past, but a valuable ally in achieving a better understanding of the world, because it leads us to think of new things and new points of view which we would not otherwise have considered.

Reduction applied to human thinking

Is it possible to take a similar, positive attitude toward reduction, regarding it as an ally rather than as an adversary, in other areas of scientific research which seem to be much more relevant to religious belief than are the laws of thermodynamics? I believe the answer is "yes," but to avoid misunderstanding, let me add that I am *not* suggesting that every phenomenological belief will survive scientific advance. Energy and entropy have been subjected to careful scrutiny and have survived intact. But phlogiston and "heat," regarded as a sort of fluid, have suffered the same fate as the geocentric universe. Thus I am not proposing that all religious belief will emerge unscathed from the critical scrutiny to which it will be exposed in a program of scientific reduction, whether in psychology, sociology or evolutionary biology. But I also see no reason to think that those beliefs which I, as a Christian, consider most central to my faith are in any particular danger of disintegrating through this process. To be sure, I may be wrong about this; one does not know what future science will be like until it arrives, and when it arrives, it will – if it is at all interesting – be very unlike what was expected. But that uncertainty faces the materialist as much as the theist. All we can do at present is to use the best scientific thinking currently available, and make some sort of guess.

Take, for example, the program of reducing human thinking to chemical reactions, or modeling it by means of computers, both of which are popular areas of research at the present time. Suppose they succeed in something like the way statistical mechanics has succeeded in explaining, or reducing, thermodynamic laws to atomic physics. In particular, suppose that at some future date science has progressed to the point where the processes involving neurons which occur inside the brain have been successfully modeled as a type of (massively parallel) computation, and suppose that the connection between these processes and human thought (consciousness, decision making, etc.) is well understood. Would it follow that man is "nothing but" a machine?

Those who have read Donald MacKay will be aware of his strenuous objection to this sort of "nothing-buttery," as he called it. I share his objections, although I prefer a different terminology. The problem with such "nothing but" statements, in my opinion, is that they take what is, or at least what could be, a valuable insight obtained from a program of scientific reduction – that the human body or brain can in some respects be usefully thought of as a machine (or like a machine) – and then claim that this is the whole story, with nothing more to be said. To illustrate what is wrong with this approach, it helps to back off from a direct discussion of human beings, and instead consider an analogous situation with fewer philosophical and emotional overtones.

Let us assume that the work-station on my desk is solving a differential equation using a fourth-order Runge–Kutta procedure. At least, that is how I think of it; let us call this a "functional description." I of course agree that inside the machine there is a rapid succession of electrical impulses which follow certain deterministic laws known to the engineers who designed it. Let us call this the "electrical" description. Is what is going on in the computer "nothing but" what is contained in the electrical description? Can the functional description be reduced to or replaced by the electrical description? More generally, how are the functional and electrical descriptions related to each other?

Unless there is some counterpart in the electrical description for what in my functional description is the multiplication or addition of two numbers in the Runge–Kutta scheme, the result which emerges at the end of the calculation is not likely to be of much use to me. In this sense there must at least be a certain *compatibility* between the two descriptions. But their actual relationship is rather complicated. One could not derive the electrical description from the functional description without a great deal of information about how the machine is put together, what programing language and which compiler was used, what initial conditions were supposed, etc.

What about the reverse relationship? Can it be said that process of solving a differential equation is "nothing but" a series of electrical impulses in the work-station? That might be the right thing to say to someone who asserts that there is, in addition to (or instead of) the electrical circuits, a little immaterial wizard inside the machine who is actually doing the work. But I think it misleading if "nothing but" is taken to mean that the functional description can simply be discarded, perhaps should be discarded, and replaced with the electrical description. As in: "While you may have thought there was a burglar in the house, what you heard was nothing but the cat looking for a midnight snack." It would, in fact, be extremely difficult, simply given the sequence of electrical states, to deduce that the machine was solving a differential equation, much less which numerical method was being employed. To be sure, some super-intelligence might succeed in doing this, but it would require considerable knowledge about the methods human beings typically use to solve mathematical problems, and a certain amount of detective work would be involved. Indeed, the most efficient method (should super-intelligences have to worry about efficiency) might be getting a hold of, or perhaps reconstructing, the symbolic form (in FORTRAN, C, or whatever) of the original program, and figuring out what I was trying to do. But we are now perilously close to the functional description which was supposed to be inessential, given the electrical description!

There may be flaws in this example, but I am convinced the basic idea is a sound one. Human understanding of phenomena in the natural world is in practice based upon a variety of descriptions, which tend to complement one another, and it is a mistake to suppose that because the same object or sequence of events can be described in several different ways, there must be one description which is "more fundamental" in the sense that all the other descriptions can be derived from it. In practice we find that this is generally not the case, and claims that a particular way of approaching the subject provides "in principle" everything we need to know should be treated with skepticism, as in the case involving entropy and atomic physics discussed earlier.

This is not to claim that the program of studying human thinking by relating it to chemical processes taking place in neurons (or something of that sort) could not lead, potentially at least, to important philosophical and religious consequences. I myself certainly hope it will do so, for that would make the research, which is obviously going to take an enormous amount of costly effort, much more interesting and worthwhile. What these consequences will be is of course very hard to say at present, when the research has not yet been done. But I see no reason to suppose that the application of methods of scientific reduction to the human body and brain will lead to the conclusion that humans are "nothing but" machines. Perhaps another example, this one very much at the center of what we think is essential to humanness, will be helpful.

Consider human freedom: the ability to choose to do something, rather than another. Should brain research eventually demonstrate that human thinking is deterministic in a manner analogous to that in which computers are deterministic, is there some way in which this freedom, which seems fundamental to our notions of moral responsibility, can be maintained as a true piece of phenomenology? Or must we resign ourselves to the idea that even though we may still think (in our less reflective moments) that we are free, this freedom is in fact an illusion? I find it helpful to address these questions by drawing an analogy with the reduction of thermodynamics to atomic physics discussed earlier. If we suppose for the sake of argument that the opposite of human freedom is determinism, and that determinism is analogous to thermodynamic energy, a concept which applies to the smallest subsystems in equal degree as it does to the whole, then there is some plausibility to the idea that human beings are deterministic, and therefore (by the definition we have just assumed) not free. Modern computers are very deterministic objects; indeed, this is what makes them so useful. And from this point of view, why should neurons be different from silicon transistors?

On the other hand, if human freedom is (vaguely) analogous to entropy, some sort of "emergent" property which can be ascribed to certain systems of appropriate complexity under certain circumstances, then its absence from computers and other sorts of un-free mechanical systems is no indication that it is an illusion. Instead, if we are going to properly understand its relationship to neurons – which, for the sake of argument, let us assume to be deterministic structures – we will need new concepts, a new approach, some new ideas which are not part of the standard theory of deterministic dynamical systems. Finding these new ideas would then be one of the things which would make this particular program of scientific reduction extremely interesting.[5]

I hardly need add that, in my opinion, the entropy analogy is better than the energy analogy, not because human freedom has any necessary connection with probabilistic models, but rather because many years of philosophical and theological controversy suggest that human freedom is not at all a simple concept, not something which, like energy, has an obvious counterpart in much simpler systems. Of course, both the energy analogy and the entropy analogy could be totally misleading, and that would be even more interesting. But in any case, the notion that human freedom will turn out to be an illusion seems to me as likely a possibility as that Pittsburgh's temperature remains above 50° F (10° C) throughout the month of February. Possible, but not something we need seriously worry about at present.

Conclusion

Is the program of scientific reduction an adversary or an ally? I, as a Christian and a physicist, consider it an ally. A great deal of scientific progress has already resulted from a program of reduction, trying to understand complicated things in terms of simpler components, and I am convinced that there is much more to be learned. We physicists may be close to the practical limits in the subatomic realm, as the cost of accelerators continues to rise. But the possibilities for better understanding how to use methods of scientific reduction remain enormous in fields outside of particle physics: in particular, in biology and psychology. It would be regrettable to have these possibilities lose public support through an irrational fear that the results will surely prove injurious to religious belief. To be sure, before the research is done one cannot say what the outcome will be. But from my perspective, the most likely casualty to further progress is not traditional theism, but instead the naive materialism which supposes that tying human thought to chemical and physical processes in the brain would somehow invalidate descriptions of human beings as free moral agents, under obligation to love God and neighbor.

Notes

1   My thinking on the matter has been heavily influenced by my interaction with the late Donald MacKay, both in person (we first met around 1971) and through his writings. MacKay started off as a physicist, but later his research interest shifted to a study of how the brain processes information, so his own professional work was in a scientific area in which reduction is one of the major research strategies. Additionally, he thought very carefully about the philosophical issues and theological implications of scientific work of this sort.
2   For a helpful overview of progress to date and problems which are still open, see Sklar (1993).
3   A note for the experts: the issue of singling out a direction of time is *not* the same as time reversal symmetry in quantum theory, which most of us believe is not an exact symmetry of nature, due to experiments on kaon decay. Instead what is relevant is that time development in classical Hamiltonian mechanics preserves the volume of phase space, and time development in quantum theory is represented by a unitary operator.
4   Mathematically speaking, (absolute) temperature is the derivative (with appropriate things held fixed) of energy with respect to entropy.
5   To be sure, these ideas may already exist. Donald MacKay, for example, was convinced that he had found the key to understanding how human freedom was perfectly compatible with a brain "as mechanical as clockwork." He did not manage to convince many other people that he was correct, but I think that his proposal deserves, at the very least, further scrutiny.

References

MacKay, Donald M. (1967) *Freedom of Action in a Mechanistic Universe*, Cambridge: Cambridge University Press.
—— (1974) *The Clockwork Image*, Downers Grove, Illinois: InterVarsity Press.
Sklar, Lawrence (1993) *Physics and Chance*, Cambridge: Cambridge University Press.

8 Mitchell P. Marcus

Mitchell P. Marcus is the Chair of the Department of Computer and Information Science and RCA Professor of Artificial Intelligence at the University of Pennsylvania. His research interests include statistical and symbolic methods for automatic acquisition of linguistic structure, natural language processing, and cognitive science. He has written or contributed to several books, including, *A Theory of Syntactic Recognition for Natural Language* (MIT Press, 1980).

Interview by Gordy Slack

GS: Would you first say a few words about your own religious background?

MM: I'm Jewish and I was brought up in a fairly non-observant household. As a child I spent every day in the Conservative synagogue across the street from my house. We moved at the end of sixth grade and I went through high school belonging to a classic Reform congregation. I was very involved with the youth group there. I hit a period in my late teens when I essentially decided I didn't believe in God, but this didn't actually have much impact on what I did. I started out in college as a philosophy major, interested in ethics. That spring was the takeover of the administration building at Harvard, where I discovered that all this wonderful ethics stuff didn't work very well when applied to the real world. So I finished undergraduate with a major in linguistics. I was basically going through school asking, "What's the nature of mind?" My questions during college were about perception and beauty and mind and the connection between them, not religious questions at all.

I finished college and got married. My wife and I both started graduate school. In May of 1974, my sister-in-law sent us a note clipped from the Boston Jewish student paper saying that a group in Somerville, Massachusetts, called Havurat Shalom, which was this very close Jewish fellowship set up as a counter-cultural seminary a few years before, was looking for members. We went and talked to a couple of people and ended up joining them. A wide variety of people who were becoming serious Jewish scholars were there, lots of ex-hippies sitting on the floor studying Jewish texts. For a few years, my wife and I were very strictly observant. I went through three years of graduate school where by day I was studying artificial intelligence and at night I was taking all these courses on Jewish Law and Jewish Mysticism. That period was a foundational period, religiously, for us.

GS: This must have been a pretty seminal time for work in artificial intelligence too.

MM: Amazing things were happening at the AI Lab. It was just becoming clear that AI was very hard. When I got there, the belief was that another four or five brilliant graduate students would solve intelligence. I was there during the period when that belief collapsed. There was this remarkable intellectual richness during the day and this other remarkable richness at night.

GS: How well integrated did you feel those two worlds were?

MM: There wasn't exactly a smooth continuum between the worlds, but there wasn't any tension. I was very aware at the time that I was doing a synthetic experimental science of the nature of intelligence. Human intelligence was the one example we had. I was interested in the question of what happens in people's minds when they analyze the grammar of a sentence, which I took as what you do when you first hear a sentence. You analyze the grammar as a step toward the meaning, mechanically, automatically, and I believed that by looking at lots of clues, including the assumption that it was computationally simple, one could come up with an account of what was going on in people's heads.

During this time, I realized that the success of physics was, in a way, catastrophic. It was really very bad, because this rigorous scientific account of the material world, over the next couple of hundred years, suggested that the only thing that existed was the material world – everything else was just poetry. It was becoming quite clear to me that the informational account was a different take on the universe. It was, in fact, parallel to, and different from, the physical. It sits on top in one sense, but in another sense not. It looked like there was emerging a second formal account of the world, so that the world could be viewed through this material lens, or could be viewed through this informational lens. They were both formal and they were both real. The minute there was room for a second account, it opened up the possibility that there were other such accounts as well. Therefore, it seemed to me that computer science was the first bit of evidence in several hundred years that there was a possibility of the spiritual being more than just poetry. For me, doing science was a form of worship – understanding how the world works was pretty powerful to me. There's this great image of Moses hiding in the rock and God goes by and Moses only sees God's back, and I remember sitting in one Saturday morning Sabbath service, and saying that God's back is science.

GS: Isn't there a tradition in mystical Judaism which holds that God is actually a kind of obscurer of ultimate truths about the world? How does that fit with this notion that science is kind of a celebration, a view of God's back?

MM: I remember thinking it's kind of astounding that the physical world is so designed. You wouldn't expect it, right? <u>The world is somehow constructed so that we can understand it.</u> How can it be that the algebra of logic gives us true sentences about the world? What is the world like, such that such a poor thing as algebra allows us this property? Somehow our informational capacity matches the structure of the universe, such that theories of things are humanly knowable. The fact that it appears to be, at least at some level, is kind of amazing.

GS: I'm wondering if you can imagine science, and scientific method, ever making friends again with the notion of the world being a meaningful place, and having that be integral to the scientific effort?

MM: I don't think so. I think the moral really remains beyond our ability to formally discuss. Our view of morality is much like our view of language. We have a set of judgments of what's acceptable and what's not acceptable in terms of linguistic sentence strings. In morality, we have a set of intuitions: this is okay; this is not okay. Then we use our intellect to construct some kind of theory of that set of judgments. But why these things are foundationally right or wrong, I think we don't have any sense of yet.

I usually talk about computer science being pre-Copernican right now. The Ptolemaic system was actually a wonderful way of describing things; it's much better if you want to understand where the stars are going to be tonight when you're looking at the heavens, than this nasty Copernican system that takes us off to the edge. Until that shift happens, my bet is that we're not quite ready for the next jump. I think this is going to take a hundred or two hundred years to sort itself out, and some people think that I'm vastly pessimistic. But I view it as vastly optimistic. It's clear that this period of time is a period of enormous explosion in a lot of ways, and it is also clear at some level that the rate of progress is speeding up. Are we going to have some account of significance? I think it would be most unlikely if we didn't. Is what we're doing going to be the precursor of that? I would be amazed if it wasn't. But I still think these things take time.

Mitchell P. Marcus

Computer science, the informational and Jewish mysticism

Introduction

This essay investigates a somewhat unlikely pair of hypotheses: first, that one major strand of Jewish mysticism actually supports both the study and the practice of computer science, which I take to include both the formal mathematical and scientific study of the informational and the engineering of informational systems; second, that our current technical understanding of computation helps us clarify a key aspect of the Jewish mystical tradition. This aspect of Kabbalah takes as central the role of speech and, as I will argue, the role of the informational in God's creation of the world. This tradition extends notions that are central to normative Judaism; public Jewish daily prayer begins with the phrase "*Baruch she'amar vehayah ha'olam*"; in English, "Blessed be the One who spoke and the world came to be."

I begin with a short digression into the role of computer science in showing scientifically that the world itself is more than just material. Next, I turn to an examination of one aspect of the Kabbalah. Finally, I look at the practice of computer science as a modern realization of this aspect of the Kabbalah.

Computer science and transcending the material

It is obvious that Isaac Newton's breakthrough in using mathematics to model the physical world in the 1600s has brought us to an astounding level of both understanding and mastery of the material world. Most remarkably, Newton showed that a simple, formal model of matter and motion accounted not only for the properties of objects in our everyday surroundings, but also for the properties of celestial bodies including the moon and the planets. As a result of this stunning success, the touchstone for truly understanding any aspect of the world within much of our society has become narrowly focused on an ability to produce a formal, mathematical, quantitative account. Beyond just this, the very existence of any putative aspect of the world around us that is not subject to such an account has become quite suspect within many circles. On this view, once physics has accounted for the nature of matter, nothing else of any substance, so to speak, remains to be accounted for.

Just as monotheism had earlier posited most elegantly that a single God was behind all aspects of human experience, much of everyday experience not withstanding, Newtonian physics gave rise to a religion of materialism that elegantly posited the reduction of all human experience in the end to a single, unitary, unified account of stuff. In very much the same way that Western religious traditions opted for theologies that posit one God with perhaps many different manifestations over those that posit many gods, the rational worldview newly emerging in the West opted for a single unified account of the world around us, namely the account provided by physics, over a world which somehow co-ordinated together many underlying, disparate aspects of existence. Most crucially, on this view, physics had relegated any talk of the spiritual to just a pretty way of speaking.

The intellectual success of computer science changes all this. In the fifty years since the invention of ENIAC, it has become apparent to those of us involved in the practice of computation that the world is informational as well as physical. This, taken naively, isn't actually very controversial. What is more surprising is the strong belief of most of us trained in computation that the world of information, or more precisely, the informational aspect of the world around us, is more or less independent of the world of stuff, of the physical aspect of that same reality.

Something quite special is now emerging in our understanding of computation. Those who have studied computation, including mathematicians, logicians and computer scientists, have developed mathematical accounts of computation and the informational which are quite separate from and independent of the mathematical accounts of the material world. From the early mathematics of Turing's model of computation in the 1930s, to formal language theory, to more recent formal accounts of the semantics of programing languages and of communicating parallel processes, we are developing richer and richer mathematical accounts of the informational, and these are simply distinct from our accounts of the physical world. In short, we now study this world of information as formally and as mathematically as the world of stuff.

But if this is true – and I believe the evidence is overwhelming – then the result of this intellectual revolution won't be merely that we can add a formal informational understanding of the world to our physical understanding. We are being invited, for the first time in several hundred years, to give up the view that a single point of reference on the universe is called for by rigorous thought. This, of course, opens up the possibility – perhaps the likelihood – that the buzzing reality around us has other independently describable aspects, including perhaps the spiritual, whose existence may one day be intellectually defended with the most formal intellectual tools we have. This is far from certain, but the emerging science of the informational provides a case study which suggests that it might just be possible.

"Blessed be the One who spoke and the world came to be"

As a computer scientist, I see the informational all around me, and I see the informational as very separate from the physical. In this section, I want to give the reader some sense of the ways in which I, as a Jew who takes my tradition seriously and who is deeply interested in the rich traditions of Jewish mysticism, see the informational as suffusing Judaism. In this section, I trace out an aspect of my religious tradition that actually supports my practice of computer science by affirming some of the key ontological underpinnings of computer science. According to this strand of the Jewish tradition, creating and animating by the manipulation of symbols, which is what computer scientists both do and study, is but one aspect of our being created in God's image.

Fundamental to the Jewish story of how the world came to be and how the world continues to be supported in its existence, instant by instant, is the role of God's speech. While the role of language in the Jewish tradition is quite central, the Kabbalah goes well beyond this in taking letters to be the fundamental units of creation, combinations of which form the words of God's speech that gives rise to the physical world.

First, the role of speech. In the Torah, the five Books of Moses, God creates through speech; "and God said let there be light, and there was light." For the God of Genesis, speech is a powerful form of action; the first utterances ever, according to the Book of Genesis, were the performative speech acts by which the world was created. Within the Jewish tradition, God's use of speech to create is central. As I noted above, in the traditional Jewish prayer book public daily worship begins with words that can be translated into English as "Blessed be the One who spoke and the world came to be." We Jews begin prayer by characterizing God as One who speaks, who creates and fundamentally uses speech to do this creation.

But the Jewish tradition goes well beyond this. To put what follows into context, one needs to know both that the tradition views the Torah as much more than just narrative text, and that the tradition has many different and often conflicting views about the nature of the Torah. One significant strand of the Jewish mystic tradition views the Torah as only secondarily a narrative text; it views the Torah as a kind of supernal DNA sequence, as it were, which encodes the entire revealed and unrevealed universe, including all of the physical, spiritual and the supernal. The sequence of letters which just happen to spell out the narrative are like the base-pair sequences of DNA; this sequence encodes everything that ever will be. Within this tradition, which appears to go back to somewhere between the second and ninth centuries, the letters of the Hebrew alphabet are taken to be the fundamental building blocks of all of creation.

Consider first a short snippet from the Midrash, the body of rabbinic accounts and commentary on the narrative of the Torah. The Midrash begins by filling in lacunae and explaining apparent inconsistencies in the text. But it goes far beyond this; it often reads back into the text the early rabbis' spiritual understanding of the world around them, sometimes unabashedly and radically revising what appears to be the plain sense of the text. From this point of view, what I relate here can be viewed as my own modern midrash on the text, an account of the text that builds on the tradition, but which is clearly the reading of a computer scientist and a computational linguist in the late twentieth century.

According to one midrash, two thousand years before God began to create the physical world, God created the Torah; God wrote it with black fire on white fire (Ginzberg 1909: 3). According to this midrash, the Torah text exists well before the world itself; before God begins to create the physical, God creates the informational. Here, the Torah text has a clear existence independent both of the physical world and of the mind of humans. This midrash says that text has a status which transcends both the physical and the psychological. Note both the similarity and the difference of this midrash to the beginning of the Book of John: "In the beginning was the Word and the Word was with God and the Word was God." Setting aside the rich contexts of the concepts of Torah and Logos, note that in this midrash, the Torah is quite distinct from God, but it is also clearly not physical.

One of the earliest books of Jewish mysticism, the *Sefer Yetsirah* (the Book of Creation), which may well date back to the second century, goes well beyond this. The *Sefer Yetsirah* is a mystical account and examination of the process of creation within the Godhead which led to this world. While much about the *Sefer Yetsirah* is quite obscure, it is clear that it views the world as created through the interaction of ten sefirot, supernal mystical aspects of God, and the twenty-two letters of the Hebrew alphabet. While I will not speak further of the sefirot here, it is clear that they are quite distinct from the well-known sefirot in fourteenth-century Spanish Kabbalah; here the nature of the sefirot appears to connect to a view of the physical embodiment of God. As the *Sefer Yetsirah* says very early in the text:

> There are ten intangible sefirot and twenty-two letters as a foundation … there are ten intangible sefirot: the number of the ten fingers, five opposite five, and in the center is set the covenant of the Only One with the word of the tongue and with the covenant of nakedness [i.e. circumcision].
> (translation from Blumenthal 1978: 15)

According to the *Sefer Yetsirah*, the letters of the Hebrew alphabet themselves are the fundamental building blocks of creation, whose combination gives rise to the totality of the world. It goes on:

> Twenty-two letters are the foundation: He engraved them, He hewed them out, He combined them, He weighed them, and He set them at opposites, and God formed through them everything that is formed and everything that is destined to be formed.
> (translation from Blumenthal 1978: 21)

It seems not unreasonable to interpret these texts together as saying that the Torah is the string of letters which fundamentally encodes in its informational structure the structure of all reality. In creating the world, God somehow translates this informational structure into the reality around us.

This tradition continued to evolve over the centuries. In the eighteenth century, one of the teachings of the early masters of the Hasidic movement was that our language and the supernal language through which the world was created are deeply connected in that the language of our prayer connects back to the supernal through the letters which make up the words or prayer. So, for example, the Baal Shem Tov, the founder of Hasidism, wrote to his brother-in-law in 1752:

> You must have the intention of unifying a name whenever your mouth utters a word at the time you pray and study the Torah. In every letter there are worlds, souls and divinity, which ascend to become bound one to the other. Then the letters become bound and united to each other to form a word and they become united with a true unification of the divine.
>
> (translation from Jacobs 1973: 75)

Within this strand of the Jewish tradition, an individual's prayer resonates with the transcendent, informational structure that existed before the creation of the physical world.

"Computer scientists are the Kabbalists of today"

Combining the notions that God creates through combinations of letters and that we ourselves are created in God's image, this Jewish mystic tradition goes one crucial step past this resonance with God's creation; it asserts that we ourselves are able to create, although in a diminished way, by combining letters. In fact, for the Kabbalists this tradition was not only an investigation of how the world was created, but also a guide to the process of creation itself. The "practical Kabbalah" taught that the utterance by humans of the correct combinations of letters done with the correct intention could somehow animate the inanimate.

One high point of the practical Kabbalah was the creation of golems, powerful but mute androids shaped of clay and animated by the appropriate incantations; the first mention of the creation of a golem goes back to the Talmud itself. According to later traditions, these incantations consisted of mathematical permutations and combinations of the letters that made up various mystical games of God. A crucial point about golems is that they are mute; by depriving them of the power of language, the tradition insists that golems themselves cannot in turn create something in their image. In this way, the Jewish tradition captures the idea that we are created in God's image, although tremendously diminished, but still with some aspect of the creative potential that characterizes God.

But of course, practical Kabbalah as the animation of the inanimate by the correct sequences of letters is just what we computer scientists do. Without the correct programing, our computers are just inanimate, inert objects. By virtue of the correct programs, of the correct sequences of letters and symbols, we animate these machines and make them do our will.

The connection between this use of practical Kabbalah to create golems and computer programing was well recognized when the first computer in Israel was named "Golem," as Gershom Sholem pointed out years ago. But a much more powerful connection emerges in a story about the early days of the MIT Artificial Intelligence Laboratory which I was told as a graduate student in that laboratory in the 1970s. According to this story, three of the most influential of the early circle of artificial intelligence researchers at MIT were told by their grandfathers that they were descendants of Rabbi Judah Loew of Prague, the famous Maharal of Prague, whose commentaries are now studied over the Internet. He was also the Kabbalist who created the most famous of the golems.

In the traditional story of the creation of this golem, the Maharal, assisted by his sons-in-law and after an appropriate period of fasting and meditation, shaped a huge figure out of river clay and brought it to life after the recital of appropriate incantations and the engraving of the Hebrew word *emet* (truth) on its forehead. After helping to protect the ghetto of Prague from a pogrom triggered by a blood libel, the Golem began to go berserk. According to the standard version of the story, the Golem was then returned to dust by the Maharal, who erased the initial letter of the Hebrew word *emet* on the forehead of the Golem, leaving the Hebrew word *met* (dead). An alternate version of the story adds that the Golem's inanimate form was left in the attic of the AlteNeu synagogue in Prague.

What follows below is the variant of this story told to me at the AI Lab in the 1970s. I believe I heard this story from Alan Brown, who was a student at the lab from the late 1960s into the early 1970s. If my memory is correct, he himself was a direct participant.

This story occurs sometime in the late 1960s before the MIT AI Lab separated from Project MAC. The AI group still consisted of a relatively small handful of graduate students, among whom were Gerry Sussman and Joel Moses, both of whom are now senior faculty at MIT; Moses is now Provost of MIT. Evidently someone noted that the first Israeli computer was called "Golem" and someone else who wasn't aware of the above story asked why. As would be expected, this led to a telling of the story of the Golem related above. But then, according to the story I was told, things took an unexpected turn. After a hesitant pause, either Sussman or Moses – I don't know which, let's say it was Moses – said that there was yet another version of the story. He said that at his bar mitzvah, his grandfather called him aside and told him that he, the grandfather, was a descendant of the Maharal of Prague, the creator of the Golem, and therefore that he, Joel Moses, was also a descendant of the Maharal. His grandfather then told him that the Golem had not been returned to dust, as in the standard story, but that the Maharal had actually put the Golem into a state of suspended animation. The Golem could be brought back to life by an incantation if the need were to arise, and it was the grandfather's duty to pass on that incantation to his grandson now that he was bar mitzvah. It would be Joel Moses' duty to pass on the incantation to his grandsons once they reached the age of religious maturity.

Not surprisingly, most of those present heard this tale with more than a little amazement. But then the other of Sussman and Moses began to speak. To the utter amazement of those assembled, he too said that at his bar mitzvah, he was called aside by his grandfather. And he too was told that he was a descendant of the Maharal of Prague, and that the Golem was in a state of suspended animation in the attic of the synagogue in Prague, and that there was an incantation to awaken it if the need arose, and it was his duty to pass this incantation on to his grandsons.

With those present looking on in complete amazement, Sussman and Moses each wrote down the incantation that he had been told, and each passed the slip of paper to the other. Evidently the two incantations were the same. At just this point, Marvin Minsky, one of the fathers of artificial intelligence and the head of the group, walked out of his office to see his graduate students looking completely astounded. He asked what had happened, and was told what had just taken place. Marvin, a complete rationalist, scoffed. "You believe this? Look, right after my bar mitzvah, I was told the same thing by my grandfather. But you think I believed it?" Minsky, evidently, not only didn't believe the story, but had forgotten the incantation, so we don't know whether his version was also the same.

Now Moses' Ph.D. thesis, finished after this episode, was a computer system called Sin, the first really powerful system for doing abstract algebra by computer; Moses' work led to the Macsyma system, the forerunner of today's Mathematica and Maple systems which perform the operations of algebra and calculus like yesterday's calculators did arithmetic. Moses dedicated his thesis "to the descendants of Rabbi Judah Loew, the Maharal of Prague."

Sussman's Ph.D. thesis was a computer system called Hacker, which wrote computer programs to solve specific tasks by adapting computer programs in its library. The dedication of Sussman's thesis reads roughly as follows: "To the Maharal of Prague, who was the first to realize that the statement 'God created man in His own image' is recursive."[1] The point of Sussman's dedication is that just as God created humankind in God's own image but somewhat diminished (where God is n, and we are $n - 1$, so to speak), so we ourselves can create beings in our own image but somewhat diminished. And, of course, what Sussman was noting in this dedication was that he had in fact created a program which wrote programs in just the same way that he believed he himself wrote them.

Soon after I heard this story, I asked Sussman and Moses directly if this story were true. Joel Moses' answer was that yes, the story was true, but that this version of the Maharal story was quite common in a certain part of the

Ukraine, and that many Jews who came from this part of the Ukraine were also told the same story about being descendants of the Maharal. He then noted, as if to prove his point, that many other computer scientists were told that they too were descendants of the Maharal, and then recited to me a short list of examples. I was too flabbergasted to remember the list, but I believe it included the likes of Norbert Wiener and John von Neumann, although I must confess I don't actually know if either of these men was Jewish. (Over the years, I have told this story to many scholars of Jewish folklore; not one of them had previously heard of it.) When I asked Sussman about the truth of this story, his response was rather different. Gerry, who I knew quite well, simply began to yell, and I made a quick retreat. Some months later, while Gerry and I were walking together at MIT talking about some bit of computer science, he suddenly changed the subject and said "You know, we computer scientists are really the Kabbalists of today. We animate these inanimate machines by getting strings of symbols just right." And then he went back to talking about our previous topic.

So what do we make of this story? Of what significance is it that three (Moses, Sussman and Minsky) of this small, early circle of AI researchers had been told that they were descendants of the Maharal? For these three, it just might be both that the story is true and that the practical Kabbalah is in their blood; perhaps they were drawn to the study of artificial intelligence as a modern form of practical Kabbalah by genetic disposition. Perhaps. But more likely, it highlights the fact that if one is steeped in the premise that we as human beings can share in some way in God's creative potential through the manipulation of abstract symbols, then one is naturally drawn to computer science and even to the pursuit of artificial intelligence. Indeed, this story itself is an invitation to enter into the modern practical Kabbalah of computer science and of artificial intelligence.

Most computer scientists understand that they perform a very deep act of creation when they program. It is fundamental to our understanding of what we do as programers that in programing we create informational structure out of the careful manipulation of symbols. I believe something stronger: that most of us understand, either implicitly or explicitly, that in programing our machines through language we emulate God's creation of the world perhaps more directly than most other forms of human endeavor. In this sense, we continue the work of the practical Kabbalists in that we continually remind ourselves through our work that we are created in the image of God the Creator, the One who created through speech. By working to actualize one key aspect of the potential resulting from our being created in God's image, our actions are a form of worship. The great modern Talmud commentator Adin Steinsaltz (1989) makes the point in the title of his book, *The Sustaining Utterance*, that God's speech is continuous, that creation is ongoing, and that if God's speech stopped for but a second, the world would disappear. Our creation of the informational is a muted echo of this ongoing creation. It is one special way in which we emulate God.

Note

1   A recursive function *f*, very roughly speaking, is a function whose value for some number *n* is defined as some combination of that number *n* and the function *f* applied to some smaller number, typically *n* − *1*. So, for example, consider the function "factorial of *n*," written *n*!, defined as *n*! = *n* × (*n* − 1) × (*n* − 2) × ... × 2 × 1, if *n* is not 0, and 0! = 1. So 4! = 4 × 3 × 2 × 1. However, for the cases where *n* is not equal to 0, we can define the function recursively as *n*! = *n* × (*n* − 1)!, if *n* is not 0, and 0! = 1.

References

Blumenthal, David R. (1978) *Understanding Jewish Mysticism*, New York: Ktav Publishing House, Inc.

Ginzberg, Louis (1909) *The Legends of the Jews*, vol. 1, Philadelphia: The Jewish Publication Society.

Jacobs, Louis (1973) *Hasidic Prayer*, New York: Schocken Books.

Steinsaltz, Adin (1989) *The Sustaining Utterance*, translated by Yehuda Hanegbi, Northvale, N.J.: Jason Aronson, Inc.

9　Bruno Guiderdoni

Bruno Guiderdoni is an astrophysicist at the Paris Institute of Astrophysics. His main research field is in galaxy formation and evolution. He has published more than fifty papers and has organized several international conferences on these issues. He is an expert on Islam in France, produces a French television program called "Knowing Islam," and has published more than fifteen papers on Islamic theology and mystics.

Interview by Philip Clayton

PC: Would you be willing to speak about the spiritual odyssey that brought you to your conversion to Islam? What were your earlier religious influences, and how did you finally come to make your commitment to Islam?

BG: My father and my mother are Christians but I was not raised in a religion. As I studied science I found that something was missing in the scientific approach to the world. I looked for other kinds of knowledge and I became aware that my quest was a religious quest. I became a Muslim ten years ago after a long spiritual path, my own readings and reflections on the nature of knowledge and the significance of human life, and a stay of two years in Morocco. I was very attracted by Eastern religions, and especially by their emphasis on the pursuit of knowledge but becoming a Buddhist, or a Taoist or a Hindu stepped too far from my experience. Becoming a Muslim is something between East and West. Islam presents itself as a religion between Western religions, Judaism and Christianity, and Eastern religions. Islamic Mysticism, which is called Sufism, puts an emphasis on the realization of knowledge in a compassionate way, within the framework of monotheism, with theological concepts which are very familiar to us. The vision of humanity, of the world, and of Creation, are very similar in Islam and in Judaism, and Christianity. Ultimately, the aim of a religious life is knowledge. This is very important because I think that my scientific quest and my religious quest both share the same quest for knowledge.

PC: As you did your graduate work in science and became a practicing astrophysicist, you recognized that there was something missing, something that the scientific pursuit of knowledge could not provide. Could you talk a little bit about what you could not get from the pursuit of scientific knowledge?

BG: I think that modern science is very successful at answering the question of how things occur. But modern science is not able to answer the question "Why are things as they are?" I think that science focuses on its main aim, the exploration of the world, and has less to say at the level of philosophy. This is probably why it is necessary to go elsewhere to find answers to questions of why. The answers to our questions give rise to other questions. Science is like an endless story, which is very exciting. But unfortunately human beings are limited in time, and we want definitive answers to our questions. This is why I was not completely satisfied by my scientific practice. I had to look to other ways of getting knowledge, and particularly to the religious path.

PC: How do you now see your work in astrophysics and Islam: as complementary, as integrated, or as very different spheres of your own self?

BG: I think they are complementary. Islam strongly emphasizes the importance of knowledge in life in general and in religious life in particular. The pursuit of knowledge is part of the faith. "Look for knowledge from birth to death," is a saying of the Prophet. The root of sin is ignorance, obscurity. So it is necessary to look for knowledge, all kinds of knowledge: the knowledge of this world and the knowledge of the hereafter. There has been no Galileo's case in Islam, no crisis between science and religion. Islam is open to all kinds of knowledge. As a Muslim, I feel very comfortable in my scientific activity because I can interpret my research work as the pursuit of knowledge of this world, as the exploration of the richness and beauty of God's Creation.

PC: It sounds as if there is a requirement to integrate your work as a scientist with your practice of Islam.

BG: Yes. Islam emphasizes God's unity. As God is One, humanity also has to be one, so any separation between the professional activity and the religious quest is not good.

PC: It has been basic to Islamic belief that Allah not only is Creator, but also is the providential ruler and the controller of the universe. Do you see that this understanding of Allah in the Islamic tradition has grown through modern science, or remained unaffected by it, or has there been a challenge to the belief and the activity of Allah through our growing knowledge of the physical world?

BG: This is a very rich question and there are several answers according to the period of history we want to focus on. There is not much debate now between science and Islamic theology because Islamic theology has almost disappeared. We are in a difficult period in the Islamic world. There are basically two great trends in Islamic thought now: one can be called the rationalistic trend, or modernism, which accepts the results of modern science without criticism. This is a global acceptance of all the results of modern thought and technology, without any attempt to see whether they are consistent with Islamic thought or Islamic theology, or if there are problems. The other trend is the fundamentalist point of view, according to which everything that comes from Western civilization is bad simply because it is Western. The fundamentalists want to elaborate an Islamic science which would be parallel to modern science, which is here considered only as Western science or Christian science. This is a complete fallacy. Moreover, this is completely contrary to the great intellectual and spiritual tradition of Islam, which insists that there is only one way of thinking. Unfortunately the debate in the modern Islamic world is very poor because there should be a third path between these two extreme ways of seeing the things.

PC: This third area that you hold to be most important would be possible only if there were Islamic theology?

BG: That's right. Unfortunately, most Muslim thinkers now are more interested in social matters than in fundamental matters because Islamic countries have to face many economic and social problems. The general trend of Islamic theological thought is to address fundamental matters. We cannot build a reflection on social problems or economic problems if there is not a reflection on fundamental matters first. This is a great weakness of modern Islamic thought and this is the reason why the results of this reflection on social and economic problems are very frequently inadequate and useless. It sometimes leads to the kind of violence that unfortunately we are seeing. Islamic philosophy almost disappeared at the end of the Middle Ages, except in the field of the mystics where reflection has always been present but rather hidden. Unfortunately there is not much contemporary reflection on these problems. It's not very easy to find a good book, or a good person who has reflected about the question you asked.

PC: Do you understand your own work as being an effort to help move Islamic theology to think about this question?

BG: I think that here, in Europe, we are more prepared to address these kind of questions because, perhaps, we have the intellectual foundations. Maybe we are more prepared to think in a quiet way because we don't have to face these problems of economy and politics that the Muslim countries are facing. We have the opportunity to have debates. I frequently go into mosques to give lectures on these problems and I see many young Muslims who have grown up here in Europe and received Western teachings and Western culture. They are waiting for this kind of reflection.

PC: So if we begin with the caveat that Islamic thinking in this area is at a very early stage, and that the answer must be very speculative from a Muslim perspective, what is your own thinking about the relation of Allah's activity in the world, and the physical description of reality which seems to leave so little place for divine presence?

BG: Everything is full of regularities and order, because God Himself is Beauty. And the beauty that we see in the cosmos is an image of God's Beauty. This is the first point. The second point is that it's not surprising that these regularities are fundamentally intellectual. They are shaped by an Intelligence who also created our intelligence so it is not surprising that we are able to explore the cosmos because the cosmos is not a foreign country. It is not different from us. The cosmos and ourselves have been created by the same Intelligence.

Bruno Guiderdoni

The Islamic worldview and modern cosmology

Introduction

In recent decades our understanding of the cosmos has been completely changed by theoretical breakthroughs based on general relativity and quantum field theory, as well as by the spectacular development of new observational techniques in astronomy and astrophysics. The resulting view of modern cosmology has broken the old pattern of a static, mechanistic and deterministic universe based on the Newtonian paradigm. We have discovered a plethora of new phenomena and placed the quest for the origins of the spacetime continuum and cosmic structures on scientific foundations.

Apart from a few individuals working in Western countries, the Muslim world as such has not participated in this endeavor. The classical period of Islamic thinking developed its concepts in connection with the medieval, Aristotelian cosmos – also known by Jewish and Christian theologians of the time – into a cosmological paradigm which was already obsolete centuries before modern cosmology was born. How, then, might the description of the universe described by modern cosmology "fit" with the worldview elaborated by Islamic thinking? Is it possible that Islamic doctrines could hold some interest for a modern cosmologist?

In this essay, I explore these questions, first by recalling the importance of the quest for knowledge in the Islamic tradition as it stems from numerous Koranic verses and Prophetic sayings. Of course, what must primarily be sought after is the knowledge of God, but the exploration of the cosmos allows us to learn how God acts. Next, I briefly review several positions which developed in Islamic theology historically on the relationship between rational and religious knowledge. I then discuss several issues which are particularly relevant for assessing how modern cosmology can be viewed from within the Islamic tradition: the intelligibility of the world, the nature of the laws of physics, the place of humanity in the cosmos, and the surprising similarities (and differences) between the scientific and spiritual pursuits of knowledge. Throughout the essay I make special reference to the teachings of three emblematic figures of the medieval Islam: al-Ghazali, Ibn Rushd, and Ibn Arabi, noting that many of the issues they dealt with are still of interest today. I conclude by emphasizing the twofold aspect of intelligence, as well as humanity's spiritual vocation.

Knowledge of the cosmos in the Islamic tradition

According to the constant teaching of the Islamic tradition, faith is intimately linked to knowledge. A famous Koranic verse (15:99) prescribes: "Worship your Lord till certainty," and many Prophetic sayings (gathered in the *Prophetic Custom*, or *Sunnah*) strongly recommend the pursuit of knowledge (*'ilm*), as a "religious duty incumbent to all Muslims, male and female." Of course, this knowledge which is our aim primarily consists in knowing God. However, all sorts of knowledge which can be connected to the knowledge of God, and which help the religious and mundane life of society, are good and have to be pursued. Moreover, according to the Prophet's well-known comment after giving technical advice on agriculture that led to failure in the crops: there is actually an area of knowledge in which religion has little to say: "You are more knowledgeable [than I] in the best interests of this world of yours." He also taught his companions that he was an "expert" in sheep-farming, as all prophets were. Because Islam does not separate the intellectual aspects of life from ethical concerns, the only knowledge which should be avoided is useless knowledge, according to another Prophetic saying: "I seek refuge in God from a knowledge which has no use." Knowledge is only a means, not an end in itself. God can deceive us with a sort of knowledge that closes our eyes to the treasures of our own spiritual vocation. We then become the erring, arrogant ones mentioned in the Koran (2:102), who "learned what hurt them and did not profit them."

There is in humans a "faculty of knowing" that is described in the Koran (16:78) according to a threefold aspect: "And it is God who brought you forth from your mothers' wombs, and He appointed you for hearing, and sight, and hearts." Hearing (*as-sam'*) is our faculty of accepting and obeying the textual indications (*an-nass*), that is, the Koran and the Sunnah which are the two sources of religious knowledge; sight (*al-basar*) is our ability to ponder and reflect upon phenomena, and is closely related to the rational pursuit of knowledge; the inner vision (*al-basîrah*) symbolically located in the heart (*qalb*, pl. *qulûb*, or *fu'âd*, pl. *af'idah*) is the possibility of receiving knowledge directly from God, through spiritual unveiling. This is why the Koran (20:114) orders the Prophet to repeat: "Lord, increase my knowledge." As a consequence, the nature of knowledge is also threefold – it is religious through the study of the Holy Scriptures and the submission to their prescriptions and prohibitions, rational through the investigation of the world and reflection upon it, mystical through inner enlightenment directly granted by God to whomever He wishes among His servants.

Since all useful knowledge brings one back to God, the Koran strongly recommends pondering and meditating upon God's Creation to find traces of the Creator in the harmony of the cosmos. Hence the so-called cosmological verses (*âyât kawniyyah*) which are frequently quoted as one of the many miracles included in the Koran (88:176–78): "What, do they not consider how the camel was created, how heaven was lifted up, how the mountains were hoisted, how the earth was outstretched? Then remind them! Thou art only a reminder." Surprisingly enough, the Arabic word for "world" (*'âlam*) stems from the same root as "knowledge" (*'ilm*). Another related word is "banner" (*'alâmah*). In some sense, the world is a banner which unveils God's presence to us. Knowing is the very act of recognizing whom the banner of the world belongs to. So the exploration of the world is strongly encouraged in Islamic thought, provided the explorer is wise enough to recognize that the harmony which is present in the cosmos originates in God.

Moreover, there is not a single "world" but a plurality of "worlds." God is called the "Lord of the worlds" (*Rabb al-'âlamîn*). It is tantalizing to interpret this as a "miraculous" indication of the large number of galaxies, stars and planets existing in the cosmos described by modern cosmology, and indeed this has been done. However, this is not the main signification of this name given to God. Instead, it indicates that Creation is not limited solely to the physical level that modern science has explored with such success. There is an infinite number of levels in Creation, which are ordered on a "vertical" scale according to their proximity to God. Our physical cosmos, with its infinite space, its infinite number of galaxies, stars, and planets, is only one of the "horizontal" planes of universal existence.

Alternatively, this plurality of worlds can be symbolically described by a system of concentric spheres. At the center is the physical world, which is embedded in the psychical worlds populated by the jinns. These psychical worlds are in turn embedded in various spherical levels of intellectual realities symbolized by the angelic realms. Put differently, our physical world is defined by more restrictive conditions and modalities than the psychical worlds (for example, a thing cannot be at two different places at the same moment, and a living creature cannot be larger than a certain size on a given planet) and consequently is less "extended" across the range of possibilities. Finally, God is called the all-Encompasser (*al-Muhît*), since He is not limited by anything, as the only absolute Being (*al-wujûd al-mutlaq*). He does not depend on the worlds (*ghanî 'ani-l'alamîn*) and He encompasses everything by His Knowledge (*'ilm*) and Mercy (*rahmah*).

Faith and reason in the Islamic tradition

How does Islamic theology articulate faith and reason? Do these two sources of knowledge always give consistent answers in the field where they both apply? There are basically six different answers to this delicate issue which have been given by various schools during the development of Islamic thought. All of them are based on the fundamental tenet that, as there is only one God, there is only one Truth. However, as we shall see, the schools strongly differ on the distance that human reason can go under its own power *en route* toward this truth.

The Traditionalists. For the Traditionalists (*madhhab al-muhaddithîn*), reason has to remain silent in front of God's prescriptions and prohibitions. God is free to change the rules of nature and justice as He wants. Thus a thing is good or bad only because God has decided so. Humanity is so weak that its own efforts to reach truth necessarily fail. Traditionalists emphasize God's power rather than the intelligibility of His acts in the cosmos. So reason has only a practical purpose for everyday life, but is not able to understand God's Creation.

The Mu'tazilites.   At the end of the eighth century and the beginning of the ninth century, the Mu'tazilites (*madhhab al-mu'tazilah*) defended the notion that revelation cannot contradict reason. For them, there is a realm of intellectual and ethical laws which God enforces in Creation. God orders or forbids a thing respectively because it is good or bad. Humanity can reach the general principles of truth with its own reason which is stimulated by revelation. The religious Law (*ash-sharî' ah ad-dîniyyah*) cannot contradict the rational Law (*ash-sharî' ah al-'aqliyyah*). Thus reason has to accompany faith from the beginning of the spiritual quest to its end.

Cheap concordism.   A new position that represents a variant of the Mu'tazilites' approach has recently been developed by reformist and modernist thinkers. They place emphasis on the "scientific miracles" of the Koran, that is, its "scientific statements," which were supposedly ignored by the early generations of Muslims and only now are proved by modern science. According to this view, the Koran reveals what is known through the natural sciences. For instance, the verse (51:47) "We built the heaven with Our might, giving it a vast expanse" is interpreted as evidence of the expansion of the universe. Thus, on this view, there cannot be contradiction between reason and revelation because revelation is (also) scientific.

Literal understanding always leads to some type of "cheap concordism." My feeling is that such concordism is dangerous for faith as well as for science. For instance, medieval thinkers could easily understand the Koranic verse (23:86) "Who is the Lord of the seven heavens, and of the glorious Throne?" as an allusion to the seven planetary spheres of Aristotelian cosmology. This attitude first leads those who defend it to reject any scientific evidence which contradicts the Aristotelian model, since they believe that it is supported by revelation, and then to cast doubt on revelation itself when new, contradictory evidence appears. I think that the agreement between religion and science occurs at a much higher level than the literal meaning of the Holy Scriptures and our scientific worldview.

The Ash'arites. During the ninth century, the Ash'arite synthesis developed, and attempted to define the place of rational knowledge in the religious pursuit by marking out the field that we can validly explore with our own reason. The great Ash'arite theologian al-Ghazali (1058–1111), known in the West as Algazel, examined the relation between science, philosophy, and religion. As all his predecessors, he believed strongly that there is only one truth, and that well-guided reason cannot be in contradiction with textual indications given by the Koran and Sunnah.

In his intellectual and spiritual autobiography *The One Who Frees from Error* (*al-Munqidh min al-dalâl*), he enumerates the list of "sciences" practiced by Islamic philosophers (*al-falâsafah*) in the wake of Plato's and Aristotle's works. Among these sciences,

> arithmetic, geometry, and astronomy have no relationship whatsoever, positive or negative, to religious matters. They rather deal with issues submitted to proof, which cannot be refuted once they are known and understood.

However, al-Ghazali writes, there is a "double risk" in their practice. On the one hand, because scientists are too proud, too sure of themselves, they often venture beyond the field where reason applies, making statements on God and religious matters which contradict textual indications. On the other hand, common believers, after seeing the excesses of these scientists, are led to reject all sciences indiscriminately. Al-Ghazali condemns "those who believe they defend Islam by rejecting the philosophical sciences," and "actually cause much damage to it." He quotes a Prophetic saying: "The Sun and the Moon are two of the divine signs. They are eclipsed neither for the death nor for the birth of anybody," and then asks, "In what measure should these words lead one to reject arithmetic which compute the trajectories of the Sun and Moon, their conjunction and opposition?"

Given that there is only one Truth, how does he deal with possible contradictions between astronomy and the cosmological verses of Scripture? The situation is clear: wherever astronomy apparently contradicts textual indications, this is the fault of the astronomers who have surely made errors in their scientific works inasmuch as they have been led to erroneous conclusions. Al-Ghazali notes that one of the most serious errors defended by philosophers is "their statement about the preexistence of the cosmos and its eternity, which no Muslim has ever claimed." Finally, in his book *The Incoherence of the Philosophers* (*Tahâfut al-falâsafah*), al-Ghazali attempts to revisit the proofs given by philosophers and to demonstrate where their errors lie.

Ibn Rushd. In his book *The Decisive Treatise which Establishes the Connection between Religion and Wisdom* (*Kitâb fasli-l-maqâl wa taqrîr ma bayna-sh-sharî'ah wa-l-hikmah mina-l-ittisâl*), Ibn Rushd (1026–1098), known in the West as Averroes, examines again the issue addressed by al-Ghazali. Ibn Rushd was a judge (*qâdî*) and his text is indeed a juridical pronouncement (*fatwâ*) to establish "whether the study of philosophy and logic is allowed by the revealed Law, or condemned by it, or prescribed, either as recommended or as mandatory." He first quotes some of the many Koranic verses (59:2; 6:75) which prompt the reader to ponder upon Creation: "Reflect upon this, you that have eyes ... Will they not ponder upon the kingdom of the heavens and the earth, and all that God created?"

As the enforcement of the revealed Law requires the use of the juridic syllogism (*qiyâs shar'î*) in Islamic jurisprudence, knowing Creation and meditating upon it require the use of the rational syllogism (*qiyâs 'aqlî*), that is, the philosophers' works that Ibn Rushd prudently prefers to call "wisdom" (*hikmah*), using a Koranic term, instead of "philosophy" (*falsafah*). But this prescription from God does not apply indiscriminately to every believer. Humanity is divided into three categories: those who, respectively, recognize truth when it is presented through a demonstrative proof (*burhân*), through dialectical arguments (*jadal*), and through rhetorical presentation (*khitâb*). The Koran includes the latter two types of arguments and prescribes the practice of reason to the first category. In some sense, science is thus considered an "esoterism" restricted to a few experts, and religion an "exoterism" which brings truth in a symbolic way fitted to the bulk of human beings. He concludes,

> since this revelation [i.e. the Koran] is true and prompts us to practice rational examination (*nazhar*) which leads to the knowledge of truth, we Muslims know with certainty that examination will never contradict the teachings of the revealed text: because truth cannot contradict truth, but agrees with it and supports it.

As a consequence, Ibn Rushd explains that wherever the results of rational examination contradict the textual indications, this contradiction is only apparent and the text has to be submitted to allegorical interpretation (*ta'wîl*). Moreover, he stipulates that there is always at least another textual indication which clearly supports this interpretation.

In his book *Incoherence of the Incoherence* (*Tahâfut at-tahâfut*), Ibn Rushd examines again the positions of the philosophers criticized by al-Ghazali. As for the issue of the eternity of the world, he presents the debate in the following way. The philosophers and the Ash'arite theologians agree that there are three sorts of being: (i) being which is produced by a cause and which is preceded by time; (ii) being which is not produced by a cause and is not preceded by time; and (iii) being which is produced by a cause and is not preceded by time. All beings of the world are included in category (i). Only God is in category (ii). The world as a whole, claims Ibn Rushd, is commonly agreed to be in category (iii), provided theologians accept the Aristotelian principle that "time is the number of motion" – which they clearly did not, by the way – and that no time existed before the existence of heavens.

As a consequence, the only disagreement between Aristotelian philosophy and Ash'arism lies in the length of the time interval that elapsed since the coming of the world into existence. The Aristotelian philosophers claim it is infinite whereas Muslim theologians and Platonic philosophers claim it is finite. Ibn Rushd closes the demonstration by quoting several Koranic verses, such as "Then, turning to the sky, which was but a cloud of vapor" (41:11) which literally means that there was "something" before the Creation of the heavens, favoring the Aristotelian position. Ibn Rushd's explanation is in a way very modern. He manages to present the issue in terms of a standard model which could be widely accepted, with the same underlying physics, and he reduces the discrepancy to a single free parameter, namely the age of the universe.

Sufism.   At the end of his autobiography, al-Ghazali tells of the intellectual confusion he suffered attempting to address the following issue: How is it possible to get certainty? He found a way out of his confusion by retiring from all rational pursuits and teachings and joining groups of Sufis who opened his heart to mystical knowledge and helped him realize the meaning of the Prophetic saying: "Knowledge is a light which God throws into the heart." Sufism, the esoteric current within Islam, proposes a large corpus of metaphysical doctrines and initiating practices, such as ritual remembrance of God's names, which aims at helping one concentrate in preparation for spiritual enlightenment. Al-Ghazali quotes several Prophetic sayings that encourage this type of pursuit, such as:

> In the days of your time, God has fragrant blasts of Mercy. Address yourselves to them, so that you may be struck by one of them, never afterwards to be wretched.

The criterion which allows Sufis not to be deceived by their own illusions (nor by the Devil's tricks) is the conformity of their spiritual unveiling to the revealed Law, according to Junayd's teaching: "The knowledge of ours is delimited by the Book and the Sunnah." Whereas the Traditionalists reject reason in their spiritual pursuit, and the theologians and philosophers try to explore the connections between reason and faith, the Sufis focus their personal efforts beyond the practice of reason, relating them to other methods.

Having briefly examined the historical relations between faith and reason in Islam, I now want to explore several issues of importance for the relationship between the Islamic tradition and modern scientific cosmology. These issues highlight the importance of the Islamic worldview for interpreting the meaning of the discoveries brought forth by modern science and for criticizing reductionist claims that matter is the only fundamental reality.

The intelligibility of the world

The fundamental mystery which subtends physics and cosmology is the fact that the world is intelligible. For the Islamic tradition, this intelligibility is part of the divine plan for the world, since God, who knows everything (one of His "most beautiful names" is the Knower, or *al-'Alîm*), created both us and the world from His Intelligence. Then He put intelligence in us. By looking at the cosmos, our intelligence constantly meets His Intelligence. The Koran (35:43; 30:30) mentions the regularities which are present in the world: As "you will find no change in God's custom ... there is no change in God's creation." Clearly this does not mean that the Creation is immutable, since the Koran frequently emphasizes the many changes we see in the sky and on the earth. It does mean, however, that there is a "stability" in Creation reflecting God's immutability. Moreover, these regularities which are a consequence of God's Will can be qualified as "mathematical regularities." Several verses (55:5; see also 6:96, 10:5, and 14:33) draw the reader's attention to the numerical order that is present in the cosmos. "The Sun and the Moon [are ordered] according to an exact computation (*husbân*)." The commentators have noticed the grammatical "intensive" form of the word *husbân*, which emphasizes the meaning "computation."

According to the views of the Akbarian school, founded upon the work of Muhyi-d-din Ibn Arabi (1165–1240), nicknamed "the greatest of the masters" (*ash-Shaykh al-akbar*), Creation is God's self-disclosure to Himself through the veils and signs of the creatures. This school often quotes the famous tradition, not from the canonical books, in which God says, "I was a Treasure but was not known. So I loved to be known, and I created the creatures, and made Myself known to them." Only God has being (*wujûd*). So the things "are" not. They only show God, through their properties which are the manifestation of a given "preparedness" (*isti'dâd*) to receive some of God's qualities. In a profound way, the meaning of the fundamental tenet of Islamic faith, that is, the attestation (*ash-shahâdah*) that "there is no divinity except God" (*lâ ilâha illâ-Llâh*), is, "there is only God in being" (*laysa fî-l-wujûd siwâ-Llâh*). However, one cannot say that things are only an illusion, or nothingness, since they show God to us under their properties. This is the reason why the "oneness of Being" (*wahdat al-wujûd*) is neither monism nor pantheism (to use the categories of Western philosophy). The paradoxical nature of Creation, between existence (*wujûd*) and nothingness (*'adam*), prevents any simple statement on its nature.

Only God is present, but He discloses Himself according to different "levels." The "five Presences" (*al-hadarât al-khamsah*) are (i) the level of the unknowable Essence (*al-Hâhût*), (ii) the level of the personal God who speaks in history and reveals His names and acts (*al-Lâhût*), (iii) the level of Kingship, that is, the intellectual world of meanings and spirits (*al-Malakût*), (iv) the level of the Kingdom, which encompasses both the physical cosmos, and (v) the psychical world (*al-Mulk*), and the level which makes the conjunction of meaning and body possible, that is, the conjunction of "intellectual laws" with physical matter and "psychical shapes" (*al-Jabarût*). God's Presence in the *Jabarût* is the world where meaning takes a body, and bodies are reconnected to their meaning. This is the locus of the mystery of regularities in the cosmos. The word *Jabarût* stems from the root *jabr* which means "constraint." Through the great "miracle" of the *Jabarût*, the laws of nature (and all the other laws which are necessary for our very existence) are continuously enforced.

God and the laws of physics

How does God act in Creation? According to Ash'arite theology, God does not act by fixing the laws of physics and the initial conditions and letting the world evolve mechanistically. As a matter of fact, there is nothing like secondary causes, simply because God, as the "primary" Cause, does not cease from creating the world at each instant. "Each day some task engages Him" (Koran 55:24). In this continuous renewal of Creation (*tajdîd al-khalq*), the atoms and their accidents are created anew at each moment. The regularities which are observed in the world are not due to causal connections, but to a constant conjunction between the phenomena, which reflects a habit or custom established by God's Will. For instance, al-Ghazali recalls the experiment in which a piece of cotton is brought into contact with fire. The fact that cotton begins to burn is only the general custom prescribed by God, who can negate this custom in order "miraculously" to preserve one of His prophets (Abraham for instance) from fire.

This principle of Ash'arite theology (called "occasionalism" in Western philosophy) completely contradicts the worldview of modern science, which stipulates the existence of secondary causes. However, one is reminded, for instance, of the critical examination of causality by Hume. The negation of causality by the Islamic tradition emphasizes the metaphysical mystery of the continuous validity of the laws. "All that dwells upon the earth is evanescent" (Koran 55:23; see also 41:53) and should fall back into nothingness. But the relative permanence of cosmic phenomena is rooted in God's absolute immutability (*samadiyyah*). This is the reason why "you will not see a flaw in the Merciful's Creation. Turn up your eyes: can you detect a single fissure?" (Koran 67:3). Newton himself attributed the constant validity of the law of gravitation to God's omnipresence in space (the "*sensorium Dei*"). Moreover, Ibn Arabi defended a still more elaborated position, explaining that God, as the primary Cause, directly acts in the secondary causes, since He is the "One who causes the secondary causes" (*Musabbib al-asbâb*).

This negation of causality need not be an obstacle to our scientific investigation of the cosmos. On the contrary, it prompts us to reflect upon the way God acts. God says in the Koran (41:53): "We will show them Our signs upon the horizons and in their souls, until they clearly see that this is the truth." Indeed, God shows up at the horizons of our knowledge. In the medieval cosmos, God was believed to act on the *primum mobile* which produced all natural motions in the lower spheres. Then, in the Newtonian cosmos, God was the Engineer who started the cosmos with its laws and initial conditions, and let it turn.

As a matter of fact, the "distance" to God continuously increased during the development of modern cosmology. Let us remember that the radius of the medieval cosmos was "measured" by the Arab astronomer al-Farghani to be 120 million km, on simple assumptions regarding the properties of the planetary spheres in Ptolemaic cosmology. This was indeed the distance to God's throne in this worldview. Now some would "interpret" the standard Big Bang model by saying that God "takes place" at the horizon located fifteen billion light-years from us.

Now, it is indeed possible that quantum cosmology will evacuate the notion of an initial singularity. If the universe has no spatial and temporal boundaries, is there still a "place" for God? The answer to this question is negative, according to Stephen Hawking, who somewhat naively interprets the Big Bang in terms of Creation *ex nihilo*. But a "horizon" of sorts is still there: why are the laws of quantum physics valid? This is the current "distance" of our horizon in the world. We can push the horizon further with our theories, but there will always be one. On the other hand, we also know that God shows up within us, and that He is closer to us than our "jugular vein" (Koran 50:16).

A famous story in the Koran relates how the prophet Abraham looked for God. He belonged to the Chaldean people who used to worship stars and planets. So naturally he gazed up at the night sky, searching for a possible sign. He first saw a star, then the Moon, which raised hope in him, but eventually they disappear beneath the horizon. Abraham commented: "If my Lord does not guide me, I shall surely go astray." Then he saw the Sun and said, "That must be my God; it is the largest. But when it set, [Abraham] said to his people: I disown your idols. I will turn my face to Him who has created the heavens and the earth, and will live a righteous life. I am no idolater" (Koran 6:76–79).

In this story, we see how Abraham's intellect enlightened by God's grace leaps from a materialistic view of God as part of the world, to a metaphysical concept in which God is the condition of the world. God is the greatest (*Allâhu akbar*), not as a "thing," not even the "largest one" (*dhâlika akbar*), but because He transcends all things. God is there, upon the horizons, where the stars and planet set, and sight fails to follow the chain of causes. As God recalls in the Koran (6:75): "Thus did We show Abraham the kingship (*malakût*) of the heavens and the earth, so that he might become a firm believer." We might even consider the prophet Abraham to be the first astronomer, since he ceased to see divinities in the sky, coming to view it instead as the locus of natural phenomena at the same time he was coming to a metaphysical understanding of the way God acts in the world. In his *Decisive Treatise*, Ibn Rushd invokes Abraham's example to prompt his readers to the study of "wisdom."

At each level of the cosmos, there are always new things. God's constant renewal of Creation means the continuous appearance of new creatures. I have previously mentioned that, according to the teachings of the Akbarian doctrine, Creation is God's self-disclosure through the veil of things. Now, God is infinite and "self-disclosure never repeats itself." The Sufi Abu Talib al-Makki, quoted by Ibn Arabi, used to say "God never discloses Himself in a single form to two individuals, nor in a single form twice." So God's self-disclosure is endless. Ibn Arabi describes this as the "sanctified effusion" (*al-fayd al-muqaddas*). New creatures and events are continuously "poured" into Creation, into disclosure. Ibn Arabi comments on the famous answer given by the Prophet to the question: "Where was God before He created the world? – He came to be in a Cloud, neither above which nor below which was any air." This Cloud (*al-'amâ'*), says Ibn Arabi, symbolizes the ensemble of "possible things" (*mumkinât*) which can be manifested and out of which there is only nothingness (*'adam*). The continuous Creation is symbolized by the "Breath of the Merciful" (*nafas ar-Rahmân*) which dilates the "possible things" out of the Cloud, from their state of "ontological contraction" in non-existence (*thubut*) (they "are" only in God's knowledge) into the realm of disclosure (and, by seeming to be "other" than God, of knowledge). What appears in Creation exactly corresponds to the flow of possible things. This is why, according to al-Ghazali, "there is nothing in possibility more wondrous than what is," because what is actually reflects God's desire to be present to us. This helps us understand the Prophetic saying: "Curse not time, for God is time (*ad-dahr*)."

God's self-disclosure founds all states of being and, more specifically, our physical cosmos. By asking the question "how" in terms of temporal causality, we "project" the ontological process of God's self-disclosure onto the physical realm and describe the effusion of God in terms of Big Bang cosmology with inflation and the current expansion stage. The infinity of God (in its all-embracing meaning, what was called categorematic infinity by medieval philosophers) requires the "infinity" – or the endless character – of the world (what was called syncategorematic infinity, meaning that there is always the possibility to add something to the sequence). Eternal inflation, and even "baby universes" described by modern cosmology, correspond to the production of this infinite number of physical universes – or "patches" of the physical universe – which are produced by God's eternal self-disclosure. But this is only the spacetime "projection" of a metaphysical principle which holds for all the Presences. The appearance of "emerging properties" at all levels of complexity, and particularly the appearance of life and intelligence, is another aspect of God's continuous self-disclosure. Thus Ibn Arabi comments: "God does not become bored that you should become bored."

Humans in the cosmos

Cosmology and ethics

The end of the medieval view was partly due to a radical change in the cosmological paradigm. The "closed cosmos" burst out and became an infinite universe. Humans were thrown out of the center of the cosmos, and we became aware that we were lost on this small planet, in an endless universe with an infinite number of "worlds" which might also be populated by other creatures (as with Giordano Bruno, for instance). Blaise Pascal could understandably say that "the icy silence of these infinite spaces" frightened him. What about his spiritual vocation? Though Pascal was himself a great scientist, he disconnected faith from knowledge.

Modern cosmology tells us that we humans are at the top of a huge cosmic building. The appearance of humanity was made possible by many "coincidences" in the laws of physics and the values of the constants, which fix the properties of the cosmic and terrestrial structures. The extension of time behind us and of space around us is a necessary condition for our very existence, as the vast extensions of the deserts of sand and ice are necessary for the ecological balance of the earth. So we should not use our smallness to deny our spiritual vocation. It is not the quantity of space and time which matters, but the quality of complexity that our existence manifests.

The Islamic tradition, however, would prefer to state this reality in a different way: for it, we are not at the top, but still at the symbolic center of Creation. Since "God created Adam upon His own form," only we are able to receive all the divine qualities. This is the reason why, following the Koran, we are "God's vice-regent on earth" (*khalîfat Allâh fî-l-ard*). Our relationship with other beings is not that from the upper to the lower level (with the concomitant possibility to exploit all "inferior" beings), but that from the central to the peripheral. This should lead to humility, not to arrogance. We are responsible for Creation, and must act in it as good gardeners. So cosmology should never be separated from ethical concerns. The astronomers indeed have a moral responsibility: they must remind us of our physical origins, and prompt us to preserve the earth.

I don't think this tells us something about the absence of other forms of intelligent life in the universe, or anything about their "inferiority." As a matter of fact, I believe that God's overabundant Mercy can create many forms of intelligent life in our infinite universe. According to the Islamic tradition, humankind has thus far received (symbolically) 124,000 prophets since Adam, and 313 messengers (those who bring forth a new revealed Law, that is, a new religion). These prophets and messengers have always transmitted the same message, that is, the remembrance of the immutable religion (*ad-dîn al-qayyim*), but through languages and under revealed forms which correspond to the peoples they were sent to. "There is no change in God's words" (Koran 10:64). Intelligent extra-terrestrial life should also receive this sort of message, and be invested by God with the responsibility of Creation at its symbolic center.

The symbolic meaning of the cosmos

As a matter of fact, my feeling is that our symbolic location in the observable universe is very similar to the one described by the medieval tradition. Our galaxy is (roughly) located at the center of a fifteen billion light-year sphere that is the patch of the universe we can observe now, because of the finite speed of light. Beyond the horizon lies the mystery of our origin. This reminds me of the medieval cosmos where we stood on the earth, at the center of a (120 million km) sphere. Beyond the crystalline heaven that is responsible for diurnal motion, lay the Empyreum, the locus of God's throne, and the mystery of our origin. Around us, there are certain local motions (in the solar system, in our galaxy, in our local group) that do not fit the larger cosmological trend. But beyond a given scale (let's say 100 megaparsecs), we find unperturbed expansion which provides us with precious cosmological information. In the Middle Ages, the local (radial) motions were those of "corruptible matter." Beyond the sphere of the Moon were the unperturbed, circular cosmological motions. We can probe the past of the universe because of the finite speed of light. This exploration brings us closer to the horizon where the mystery lies. In the Middle Ages, the saints reproduced the Prophet's spiritual Journey and Ascension (*al-isrâ' wa-l-mi'râj*) through the heavenly spheres to God's throne, as Dante Alighieri did in his *Divine Comedy*.[1]

Of course, this comparison should not be taken too seriously. I only mean to suggest that our symbolic location has not changed all that much since the Middle Ages, even if our physical location is now known in greater detail. Our spiritual vocation requires the possibility of cosmology as a science. And indeed cosmology is possible, contrary to what Auguste Comte and the positivists claimed in the nineteenth century. The universe lets itself be known, but only partly. We can build a cosmology from a few elements based on common sense, or on a large amount of sophisticated observations and theories, but we always face the mystery of the origins of the cosmos.

The endless quest for knowledge

The growth of scientific knowledge

The philosophical definition of scientific truth is a debated issue. For instance, in his works, Sir Karl Popper has claimed that we cannot say that a theory is true, but that we can say that it is wrong whenever there is strong experimental or observational evidence which contradicts its predictions. Good theories have to propose crucial experimental tests and, so to speak, "stick their necks out." Popper's ideas on truth and falsifiability have also been criticized, because experiments and observations themselves are theory-laden. So, at a given stage, it can be difficult to choose whether a theory should be rejected or the experiments and observations revisited or discarded. In fact, apart from theories which are not self-consistent, it is difficult to identify the conditions for a theory to be true, probably true, wrong, or probably wrong – even if we can and must keep our everyday definitions of truth and falsehood for practical purposes.[2] But what we can define with great accuracy is the method which leads to the development of scientific knowledge. The philosophical "truth" of science lies in the fecundity of its method. Cosmology, as well as the other sciences, is fundamentally an ongoing process.

We can view this process as continuous approximation of truth, and indeed, it is sometimes so. But, at other periods of its development, science has opened completely new paths. We can also defend the idea that the scientific pursuit leads to the growth of our knowledge *and* our ignorance, since the answers to our questions immediately suggest new questions. There is at present a large debate over whether a "Theory of Everything" (TOE) is possible. Such a theory would have to predict all the laws of physics, the values of the constants in the laws, the properties of elementary particles, the initial conditions, and should also explain by itself why it is not only necessary, but unique. Gödel's theorem shows that mathematics itself suffers from incompleteness, since there are true statements which cannot be demonstrated within a given set of axioms. Perhaps the TOE could use mathematics lying in an island of completeness. However, the Islamic tradition sees the world as the locus of constantly emerging new properties. Thus, from an Islamic point of view, the elaboration of a TOE would appear to be impractical.

The apophatic nature of spiritual knowledge

There are surprising similarities between the open process which leads to the growth of scientific knowledge and the pursuit of religious knowledge articulated by the metaphysical doctrines of Islam. Here, the dogmas represent what can be said, and must be said, about our spiritual vocation as human beings. But, of course, the ideas we make about God are limited to our own understanding. They immediately become idols if we think we have gotten the truth. So we have to reach for the highest notion of God by destroying the idols we continually create. This is one of the meanings of the *takbîr*, the words "God is the greatest" (*Allâhu akbar*) which are repeated during the ritual prayers. By fighting against themselves to proceed toward a higher idea of God, believers answer God's words (in the *hadîth qudsî*): "I am with My servant's opinion of Me." As a matter of fact, the believer discovers that any statement upon God's Essence is impossible, according to the Prophetic saying: "Reflect upon all things, but reflect not upon God's Essence." This is why Abu Bakr, the first caliph, was given to saying that "incapacity to attain comprehension is itself comprehension." The growth of spiritual knowledge primarily is an increase in the level of the believer's bewilderment and perplexity (*hayrah*), who eventually realizes the meaning of the Prophet's prayer: "O God, increase my bewilderment in Thee!"

Conclusion

Thus, for a Muslim cosmologist, the exploration of the cosmos is a way of worshiping God. God's Creation of human beings makes science possible since our intelligence finds a trace of God's Intelligence in the harmony of the world. In some sense, believing scientists look for answers to the "how" question in their scientific pursuit, and to the "why" question in their spiritual quest.

The Greek and medieval philosophers (Jews, Christians and Muslims) distinguished two aspects in human intellect (Greek: *nous*; Arabic: *'aql*): (i) the faculty of pondering (Greek: *dianoia*; Arabic: *fikr, tafakkur*), which allows us to use syllogisms and validly produce true statements from other true statements; this is the algorithmic power of human mind, what we call reason; (ii) intellectual intuition, that is, the ability to grasp truth immediately (Greek: *noêsis*; Arabic: *hads*). Because of the spectacular progress in our scientific understanding of the universe, we have forgotten the latter aspect, which is equally necessary. In their spiritual commitments, scientists look for recovering this contemplative aspect of truth. In the Islamic tradition, one would say that the intellect (*'aql*) should be aware of the limits of pondering (*fikr*).

However, the spiritual pursuit is not limited to the intellectual contemplation of truth, as can be seen in the following story. Around 1180 Ibn Rushd and Ibn Arabi happened to meet in Cordoba. Ibn Rushd, who then was already a renowned philosopher, was eager to meet the young Ibn Arabi. The latter reports on their meeting:

> When I entered in upon [Ibn Rushd], he stood up out of love and respect. He embraced me and said, "Yes." I said, "Yes." His joy increased because I had understood him. Then I realized why he had rejoiced at that, so I said, "No." His joy disappeared and his color changed, and he doubted what he possessed in himself.

Then comes the explanation of these strange exchanges. Ibn Rushd asked the crucial question: "How did you find the situation in unveiling and divine effusion? Is it what rational consideration gives to us?" Ibn Arabi replied, "Yes and no. Between the yes and the no spirits fly from their matter and heads from their bodies." He reports Ibn Rushd's reaction: "His color turned pale and he began to tremble. He sat reciting, 'There is no power and no strength but in God', since he has understood my allusion."[3]

Ibn Arabi alluded to eschatology by recalling that even if reason can go far in its attempt to grasp reality, no one has been intimately changed by scientific knowledge. Knowing Pythagoras's theorem, the theory of general relativity, or Big Bang cosmology does not change our worldview – perhaps the way our mind eventually works, but not our "heart" (in its traditional meaning). These discoveries are, of course, important milestones in intellectual history. They can also produce very strong feelings in those who dedicate their lives to such study. But revelation speaks of another type of change that prepares us for the afterlife (*al-âkhirah*).

To pursue our quest for knowledge, we shall have to leave this world and enter another level of being which is a broader locus for God's self-disclosure. The Islamic tradition promises that the quest for knowledge will end when the elect contemplate God's Face on the "Dune of Musk" (*al-kathîb*) which is located on the top of the heavenly Gardens. But this end of the quest will not be the end of knowledge. On the contrary, the elect's contemplation of God will continuously be renewed, as they will know "what no eye has seen, what no ear has heard, and what has never passed into the heart of any mortal." Our reason may estimate that this is impossible, since we do not see "how" it can physically happen. But indeed, the Dune is the locus of the answers to the question "why," not "how."

Notes

1   Thanks to Miguel Asín Palacios (1926), we now understand the significant influence of the Islamic worldview on Dante's works.
2   Here I refer particularly to the bright presentation by Alan Chalmers (1999).
3   For the translation, see Chittick (1989). I also acknowledge the excellent presentation of Ibn Arabi's work given in this book.

References

Asín Palacios, Miguel (1926) *Islam and the Divine Comedy*, translated and abridged by Harold Sunderland, London: J. Murray.
Chalmers, Alan F. (1999) *What is This Thing Called Science?* 3rd edition, Indianapolis, Ind.: Hackett Pub.
Chittick, William C. (1989) *The Sufi Path of Knowledge*, Albany, NY: State University of New York.

10 Michael Ruse

Michael Ruse is Lucyle T. Werkmeister Professor of Philosophy at Florida State University, having retired from the University of Guelph in Canada, where he spent thirty-five years. A fellow of the Royal Society of Canada and the American Association for the Advancement of Science, he is a former John Simon Guggenheim fellow and an Izaak Walton Killam fellow. He is the author of *The Darwinian Revolution: Science Red in Tooth and Claw* (University of Chicago, 2nd edn, 1999), *Monad to Man: The Concept of Progress in Evolutionary Biology* (Harvard University Press, 1996), and *Can a Darwinian be a Christian? The Relationship between Science and Religion* (Cambridge University Press, 2000).

Interview by Gordy Slack

GS: Could you say a few words about your religious history?

MR: I was born in 1940, which was the first year of the Second World War in England, and I was born to lower-middle-class parents who had fairly radical Marxist beliefs. Amongst other things, they were pacifists. My father was a conscientious objector during the war, during which time he came in contact with the Quakers. After the war, my parents joined the Society of Friends. Although I'm not a birthright member, from my earliest conscious memories I was brought up as a practicing member of the Society of Friends in England.

At the age of twenty-one or twenty-two, by which time I had started doing philosophy, I had a strong reaction against it all and ended up on the atheistic end of the spectrum. Like a lot of people, I've been swinging back much more toward the middle. My natural inclinations are away from extremism of any kind. I feel uncomfortable about evangelicals who are either for or against God. I don't see myself swinging over the other way again, though I still think that there are insuperable objections to Christian belief. The question of the veracity of claims about Jesus's divinity and so forth make me very uncomfortable.

If somebody were to ask me now, I'd probably describe myself as an agnostic. But I'm always a bit wary of that, because "agnostic" so often covers people who are basically not interested in religion. I'm much more agnostic in the Thomas Henry Huxley sense, actively interested in the whole issue, but very skeptical about what's going to happen.

GS: Would you talk a little bit about the fit between Christianity and evolutionary theory?

MR: If faith is to mean anything, if the alienation from God is to mean anything, it has to be because the whole tradition of natural theology isn't going to work. Faith has to be a leap into the absurd. I find that very attractive, but I don't have faith. It's not that evolution proves that the world is meaningless. The world is meaningless and evolution is part of that. That's why my whole book, *Monad to Man*, has been on progress, because I think that people like Ed Wilson are secular natural theologians. They want to find a meaning. For me, there are no routes to God through reason.

GS: Is there an evolutionary explanation that comes to mind for this alienation between God and man?

MR: The whole question of science is an irrelevancy with respect to religion. If religion is true, science is a wonderful testament to the human spirit and to God's gift to us, and to our reflecting his intelligence. I don't see any tensions there at all. I feel much more tense about the problem of evil, for instance. And about the question of rival religions to Christianity; it does seem to me that Jesus made some fairly solid claims to uniqueness – as one reads the Bible, one finds the deep anti-Semitism of the Gospel according to St. John. Those are things which are real obstacles for my acceptance of Christian belief as opposed to science.

One of the things I'm trying to do is to find why people have so much trouble with evolution. I'm convinced it's because evolution has been pushed so hard. Some people melded their Christianity and their evolution. Others didn't. But I think it's because evolution has had this notion of progress at its heart, and progress is opposed to providence. As I wrote in *Monad to Man*, there are real tensions in trying to bring evolution and Christian belief together because evolution, with its notion of progress, can function as a religion. Progress is the idea of doing it for yourself. Providence is the idea of God doing it all for you through his divine Grace. Clearly if you accept providence, then without God you are as nothing – to think otherwise is heretical.

GS: I'm not sure exactly how the notion of purpose fits with either providence or progress, but there is a sense of meaning imbuing the lives of many practicing Christians, the conviction that things are occurring for some purpose.

MR: I am empathetic to natural supernaturalism which says that the whole experience of life is such a miracle, so wonderful, that you cannot deny the reality of something more. I take it that this is what faith is all about, at least at some level. Whether you want to interpret this in Christian terms or some other terms is another matter. I have a great respect for somebody who says that. But my skepticism is an active skepticism. I don't hate Christianity in the way that Huxley did. I don't believe in it, but it makes sense that there has to be something more. If somebody tells me of their belief in God, I envy their faith. I'm not just being snarky, but I don't have it. And if somebody tells me I should let providence make it happen, I tell them that this is not the way I do things.

GS: How do you relate the pursuit of truth in science to the pursuit of other kinds of truth, including spiritual and religious truth?

MR: I don't really. I think life's a quest. Some people are interested in science, some people aren't. I take science to be an important and fundamental way of understanding. I would say it's the best way of understanding we've got. I don't necessarily want to say that it's the only way that we've got. Poetry, music, opera, these things can speak to us in ways that science clearly doesn't. If pressed, I would say that I think that any religious beliefs must fit in with one's scientific beliefs. But as I've said, for me faith may stand outside science. Religion does not have to be reasonable in a scientific sort of way.

GS: How could these activities be carried simultaneously in human life?

MR: The point as far as I'm concerned is that you've got to start with the science. That's not to say that it is the foundation. But at a certain level, it's the necessary condition. You try to understand the world as best you can in that context. Once you've done that you determine where everything else fits in.

GS: Some of the scientists I've talked to refer to the evolution of parallel languages in their own lives and also in public discourse, particularly in American culture: the language used to talk about religion, and the language used to talk about science. These seem in many ways to be alienated worlds. But in some cases there is at least a fairly congenial relationship between the two.

MR: Part of the problem is that the metaphor of languages suggests that science and religion are the same sorts of things, and that they're co-equal, or one wins and one loses, or something like that. I am uncomfortable about that kind of metaphor. Once you start doing that, you're comparing and contrasting. If you do that, religion is going to lose. Now, at another level, we've got to use metaphors. Langdon Gilkey talks about "why" questions, and "how" questions. Perhaps that's the only way that we can handle these things. I'm uncomfortable about that kind of metaphor because I don't think religion functions that way: science can't answer it, so let religion take over. I'm at a disadvantage not having any faith myself. If I did believe, then religion for me would be a different dimension of experience. Then you would be right, I would want to say that the two approaches were not alienated, but different. Certainly there are people who have managed to find a place both for their science and for their religion. I'm not one of them, but I respect that ability.

Michael Ruse

Darwinism and atheism

A marriage made in heaven?

Introduction

In the past decade, there has been an interesting meeting of minds. One finds, as one would expect, that the Christian fundamentalists – the biblical literalists or so-called "Creationists" – have argued that Darwinism and Christianity are incompatible. This they have argued for the past century. For these Christians, every word of the Bible must be taken at immediate face value: the early chapters of Genesis tell truly of a six-day period of Creation only a few thousand years ago, of humans appearing miraculously at the final moment, and of a universal flood destroying all but a chosen few of those animals living in the earliest times. Hence, understanding by "Darwinism" the belief that all organisms living and dead have arrived by a slow process of evolution from forms very different and probably much simpler, and that the process of change was natural selection – the survival of the fittest – the incompatibility follows at once. On this version of Christianity, one simply cannot simultaneously be a true Believer and a Darwinian evolutionist. Since the fundamentalists tend to regard anyone who does not subscribe to their beliefs as denying just about anything of ontological theological worth, they therefore regard Darwinism and atheism – the absolute rejection of God's existence – to be tightly linked. Indeed, for the Creationist, Darwinism is simply atheism given a scientific face, and it has to be the main reason why anyone would reject belief in a Christian God.

What one also finds today – and this perhaps one might not expect – is that a number of articulate, prominent Darwinians agree entirely with the Creationists. They too see science and religion in open contradiction. In this essay, I shall look at this claim that Darwinism and atheism are different sides of the same coin. I shall consider what connection exists between the two. Deliberately, I shall limit my discussion inasmuch as I shall not consider the truths of either Darwinism or atheism as such. And in another way I shall limit my discussion, in that I shall not look directly at the arguments of the Creationists, the biblical literalists, on this matter.[1] Rather, the focus of this essay will be on the arguments put forward by my fellow Darwinians. Let there be no concealing of the fact that I am as ardent an enthusiast for evolution through natural selection as anyone, not excluding Dawkins himself (Ruse 1986, 1989, 1995). Nothing I have to say now qualifies or mutes that enthusiasm.[2]

Let us turn then to several of today's Darwinian evolutionists who would erect impassible barriers between their theory and the Christian religion.

Edward O. Wilson

I shall begin with the Harvard entomologist and sociobiologist Edward O. Wilson. I am aware that much of what he wants to say is highly controversial, but this is not the topic of this essay. I and others have looked at these matters at length elsewhere. Here I am assuming that the science is well taken. I should add that I am also aware that, in many respects, Wilson owes much in his thinking to Herbert Spencer, perhaps at least as much as to Charles Darwin (Ruse 1996). However, in the present context, the thinking is essentially Darwinian: I shall therefore ignore historical and conceptual niceties.

Wilson is an interesting case. Although he is no Christian, in many respects he is significantly more sympathetic to religion in general and perhaps even to Christianity in particular than many Darwinian non-believers. Here, there is a stark contrast with the two men to be considered in the following sections. Wilson recognizes the importance of religion and its widespread nature: he is very far from convinced that one will ever eliminate religious thinking from the human psyche, at least as we know it. In his popular Pulitzer-Prize-winning work, *On Human Nature*, Wilson devotes an entire chapter to religion. As far as he is concerned, religion exists purely by the grace of natural selection: those organisms which have religion survive and reproduce better than those which do not. Religion gives ethical commandments, which are important for group living; also, religion confers a kind of group cohesion – a cohesion which is a very important element of Wilson's picture of humankind.

One should note that, although Wilson talks about cultural evolution, he makes it clear that in fact he thinks that, in some sense, religion is ingrained directly into our biology (1978: 174–175). Thanks to our genes, it is part of our innate nature. What Wilson argues is that, in some sense, the biological advantage conferred by religion will remain forever. However, he does believe that giving a Darwinian explanation – he would call it: giving a "sociobiological" explanation – does make it possible to deny religion the status of a body of true claims. And indeed, given our religious needs, this means that in some sense Wilson's position requires that the biology itself become an alternative secular religion. (Wilson much admires Julian Huxley, the grandson of Thomas Henry Huxley, a leading evolutionary humanist in the first part of this century, who authored a book entitled *Religion Without Revelation*.)

Although I am not interested here in critiquing Wilson's scientific position, I would agree with critics that Wilson's view seems to be rooted in his own childhood experiences of Baptist fundamentalism in the American south, as much as in any knowledge or study of empirical reality. But let us take his position at face value and ask what Wilson's view implies for Christianity, particularly *vis-à-vis* the whole issue of atheism.

I take it that, in Wilson's own mind, what is happening is that Darwinism is explaining religion (including Christianity) as a kind of illusion: an illusion which is necessary for efficient survival and reproduction. Once this explanation has been put in place and exposed, one can see that Christianity has no reflection in reality. In other words, epistemologically one ought to be an atheist. As I have said, what makes Wilson particularly interesting is that – atheist although he may be – he still sees an emotive and social power in religion. He would therefore replace spiritual religion with some kind of secular religion – which, it turns out, happens to be Darwinian evolutionism.

Of course, the kind of argument that Wilson is promoting is hardly new. Both Karl Marx and Sigmund Freud proposed similar sorts of arguments: trying to offer a naturalistic explanation of religion, arguing that once one has this explanation in place, one can see that the belief system is false. So already I doubt the absolutely essential Darwinian component to the general form of the argument. But, this apart, is the inference in general well taken? And even if it is well taken, what of the specific case of Darwinism and Christianity?

First the general case. It is certainly true that, sometimes, an explanation of why someone holds a belief suggests that, with respect to truth, the belief is not particularly well taken. Consider for instance the instance of spiritualism, particularly as it pertained to people's beliefs and practices in the First World War. Many bereaved people turned to spiritualism for comfort: indeed, they derived such comfort, for they heard or otherwise received messages from the departed. However, I suspect that all of us would agree that, even in those cases where no outright fraud was involved, it was unlikely that the dead soldier was in fact speaking to those remaining. People's strong psychological desires to hear something comforting led them to project and receive the desired messages, and so they heard them. Once one offers this explanation, seeing how unreasonable it is to expect that the departed were in fact speaking, the whole spiritualist position collapses.

Yet, not all explanations of why we or how we come to believe things are necessarily such as to debunk the veracity of the belief systems. Suppose, for instance, one gives a physiological/optical explanation of sight, showing how it is that someone is able to spot a speeding train bearing down on them. The fact that one can give an explanation – in terms of the eye's physiology and of light rays and so forth – in no sense demotes or discredits their belief that a speeding train is indeed bearing down on them.

The question we must ask now is whether religion is more like the spiritualism case or more like the speeding train case – and it is surely pertinent to note that this is a question which is neither asked nor answered by Wilson.

This omission does not mean that Wilson's preferred option for religion – spiritualism rather than train – is wrong. But it is to say that some additional argument is needed, in order to show that religion is more like the spiritualism case than the speeding train case. The point I am making is that, in a way, arguments like that of Wilson – as indeed like those of Marx and Freud before him – are arguments that, to a certain extent, come after the event rather than before. One realizes religion, let us say Christianity, is in some sense inadequate or false. Then, one is led to ask why exactly it is that people are led to believe it and one offers some kind of materialistic or naturalistic argument to explain it. I am not sure that the explanation in itself is sufficient to show that one's belief is false, or at least I think one needs some further information as to why the explanation itself shows the belief false.

This brings us to the particular Wilsonian case of Darwinism and Christianity. And here the missing elements in Wilson's case become crucial. The fact that one has an evolutionary explanation of religion is surely not in itself enough to dismiss the belief system as illusory or false. We might offer an evolutionary explanation as to why somebody spots a speeding train, but the fact that it is an evolutionary explanation does not make the existence of the speeding train fictitious. Indeed, if anything, the evolutionary explanation convinces us that we do have a true perception of the speeding train. If evolution led us to think that it was a turtle dove rather than a train it would not be of much survival value.

None of this is to deny that people have proposed arguments suggesting that belief in Christianity is unsound, even ridiculous. There are all sorts of paradoxes which the Christian must face. But whether or not one can defend Christianity against such charges, I do not see that the charges themselves have been brought on by Darwinism: which is the nub of this discussion. Take the problem of the tension between free will and God's grace. If indeed we are free, does this mean that we can raise ourselves up in some sense? But if we can, then what need have we of God's grace? Yet, if God Himself chooses who is to be saved and who is not to be saved, where then is the bite of human freedom? I am not saying that one cannot explain away or deal with paradoxes like this, and of course we have had two thousand years of theological attempts to do precisely this. My point is that Wilson does not in any sense show that Darwinism adds to this problem. It is a paradox whether we evolved or were created miraculously on the sixth day. In short, Wilson's Darwinism does not prove the inadequacy of Christian belief; rather, his Darwinism shows why one might have a Christian belief, if evolution be true.[3]

Let's try again. Could one not argue that Darwinism shows that there is something wrong with religion, since Darwinism does not have a built-in teleological progression up to a particular end? At the least, this non-directedness shows the possibility of reaching and holding alternative religious beliefs – each one of which could do the work selection demands. But, if indeed we could as readily reach alternative religious beliefs to any that we do hold now, and if all these beliefs could be sincerely held, then this suggests that maybe any religion as such does not exist in its own right because it is true. It, like all of its possible alternatives, is simply a fiction created by biology to enable us to survive and reproduce. In this way, there is a difference between religion and the train example. It is true that different beings might – and indeed do – evolve different ways of sensing the train's approach. One uses sight, another uses hearing. But the long and the short of it is that one is going to have to sense the train in some fairly standard sort of way, otherwise one is going to be wiped out. Religion, however, might be effective in achieving group cohesion, even though it takes on very different forms. It might take on one of an infinite number of forms. All of which suggests that Darwinism tends to corrode Christian belief more than one might expect.

But, obviously, one can mount this argument, even today, without really bothering too much about evolutionary biology. We know full well that different people do have different religious beliefs. Some are Christians, some are Jews, others are Muslims, and so on. In other words, what we know already is that cultural evolution, if we can so call it, has led to different religions that people maintain sincerely. And so already we seem to have at work an example of the argument based on the non-progressivist nature of creative causal processes. Moreover, I hardly need say that there are already those today who think the argument is significant and quite corrosive of Christian belief, or indeed of any specific religious belief. I hardly need say also that there are standard replies that can be offered. One can suggest one belief is better than others, and that some people having been led to mistaken beliefs does not deny the truth of this one belief. Many people sincerely deny evolution, but this does not make them right. Or one can argue that perhaps there is some common core to all religious belief, and that this is what counts. And note that, as with the main argument, these counter-arguments have little to do with Darwinism. The point I am making is that for all the important issues here – and I am sincere in agreeing that there are – I am not sure that Darwinism is particularly relevant. Christian belief is being judged by other factors.

In any case, somewhat paradoxically, I am not sure that Wilson himself would particularly want to push this argument about the significance of the non-directedness of the evolutionary process. As I have mentioned, he is strongly influenced by Spencer. So although he clearly allows for cultural variation when it comes to biology, I suspect that Wilson would say that all evolved religion shares a common core. Any new or hypothetical evolution must be similar to what we today regard as religion: otherwise it would not maintain group cohesion, and so on. My suspicion therefore is that Wilson would be inclined to downplay this comparative argument, at least at the biological level.

But what about a pure Darwinian, who would deny any progress or direction to evolution? Could they not launch an argument of this type against the validity of religion? Perhaps they could, although even here whether one now has something which makes Christianity untenable on Darwinian grounds is another matter. Even if it is possible for people to be biologically (as opposed to just culturally) insensitive to religion, it is still open for the Christian to argue that those who did evolve this way are in some sense religiously blind, as we know some people are color blind. I am not sure (without more argumentation) that this is an entirely effective response, but it is at least a response which one could make.

All in all, therefore, although I think that Wilson's argument is important, and one that a Christian ought to take seriously, I am not convinced that Wilson shows that Darwinism implies atheism. The atheism is being smuggled in, and then given an evolutionary gloss, which is an entirely different matter.

Richard Dawkins

Let me start by quoting from an interview that Dawkins gave recently.

> I am considered by some to be a zealot. This comes partly from a passionate revulsion against fatuous religious prejudices, which I think lead to evil. As far as being a scientist is concerned, my zealotry comes from a deep concern for the truth. I'm extremely hostile towards any sort of obscurantism, pretension. ... There's a certain amount of that in religion. The universe is a difficult enough place to understand already without introducing additional mystical mysteriousness that's not actually there.
>
> (Dawkins 1995a: 85–86)

I am sure the reader will not be surprised to learn that Dawkins has recently characterized his move from religious belief to atheism as a "road to Damascus" experience (Dawkins 1997b). Saint Paul would have recognized a kindred spirit. For Dawkins, either you believe in Darwinism or you believe in God, but not both. For Dawkins there is no question of affirming what philosophers call an inclusive alternation, that is to say either *a* or *b* or possibly both.

Why then not simply slough off Christianity and ignore it? Things are not this simple: Dawkins – like any good Darwinian including Charles Darwin himself – recognizes that the Christian religion poses the important question, namely that of the design-like nature of the world (Dawkins 1986). Moreover, Dawkins believes (what I suspect many philosophers would deny) that until Charles Darwin no one had shown that the God hypothesis, that is to say, the God-as-designer hypothesis, is untenable. More particularly, Dawkins argues that until Darwin no one could avoid using the God hypotheses. (I am not sure that this is historically correct either, but no matter here.) In other writings, Dawkins deals at some length with the arguments of David Hume (1947), showing that although Hume offered a devastating critique of the argument from design (the teleological argument), ultimately he had to agree that there was something there which needed explanation. Moreover, as things stood at the time of Hume, that something by elimination had to be God.

As things stand, however, Dawkins's argument against inclusive alternation is altogether inadequate. Why should we not say, with earlier Darwinians who were also Christians, that the alternation is inclusive? Why should we not say that Dawkins is certainly right – as against say someone like Stephen Jay Gould (Gould and Lewontin 1979) – in stressing the design-like nature of the organic world, but that he is wrong in thinking that it is either Darwinism or God, but not both? At least, even if he is not wrong, he has failed to offer an argument for this, and there have been many in the past who quite happily argued that the design-like nature of the world testifies to God's existence, that God simply created through unbroken law. Indeed, people have in the past argued that the very fact that God creates through unbroken law attests to his magnificence. Such a God is much superior to a God who had to act as Paley's watchmaker would have acted, that is, through miracle.[4]

In fairness, I think that at this point Dawkins does have a second argument up his sleeve. It is the venerable argument based on the problem of evil. But for Dawkins it is more than just the traditional argument (which is in itself not particularly evolutionary). What Dawkins would argue is that not only does evolution intensify the problem of evil, but Darwinism in particular makes it an overwhelming barrier to Christian belief. This argument is expressed most clearly in one of Dawkins's recent books, *A River out of Eden*. In a chapter entitled "God's Utility Function," he writes:

> nature is not cruel, only pitilessly indifferent. This is one of the hardest lessons for humans to learn. We cannot admit that things might be neither good nor evil, neither cruel nor kind but simply callous – indifferent to all suffering, lacking all purpose.
>
> (Dawkins 1995b: 95–96)

Then, later in the chapter, Dawkins talks about organisms being excellent examples of design-like engineering. If we tried to unpack the engineering principles involved in organisms, the problems of pain and evil come to the fore. Meaning by the notion "utility function" the purpose for which an entity is apparently designed, Dawkins writes as follows:

> Cheetahs give every indication of being superbly designed for something, and it should be easy enough to reverse-engineer them and work out their utility function. They appear to be well designed to kill antelopes. The teeth, claws, eyes, nose, leg muscles, backbone and brain of a cheetah are all precisely what we should expect if God's purpose in

designing cheetahs was to maximize deaths among antelopes. ... Is
He a sadist who enjoys spectator blood sports? Is He trying to avoid
overpopulation in the mammals of Africa? Is He maneuvering to maximize David Attenborough's television ratings? These are all intelligible
utility functions that might have turned out to be true. In fact, of
course, they are all completely wrong. We now understand the single
Utility Function of life in great detail, and it is nothing like any of
those.

(Dawkins 1995b: 104–105)

Dawkins's point here is that if there be a God, then He is one who certainly
is nothing like the Christian God: He is unkind, unfair, totally indifferent.

Now let us look at the worth of these arguments of Dawkins, remembering that the question at issue is not whether atheism is a sound religious
position to take, but rather whether there is something inherent in
Darwinism which pushes one toward atheism. My suspicion is that Dawkins
still fails to make his case, although I do not want to minimize the importance of his arguments. I should say that (amusingly) my suspicion is
reinforced by the fact that Jean-Henri Fabre, whom Dawkins cites in his
own support, was in fact a Creationist opponent of Darwin who thought
that one must invoke a designer because evolution on its own is inadequate
(Tort 1996)!

Let me start by saying that I think Dawkins is absolutely right in his
belief that Darwinism does impinge on the argument from design. More
than this, at a conceptual level I would agree with Dawkins that in some
sense Darwinism does make it possible to be an intellectually fulfilled
atheist, or some such thing. I agree that before Darwin conceptually it was
difficult to see how design could be explained naturally and that design
certainly did need an explanation. I agree with Dawkins – and of course
with Darwin and Paley – that the design-like nature of the organic world is a
major problem standing in need of explanation. I agree incidentally that
Hume saw this as an insuperable objection and this (and not some false
sense of expediency) is why he equivocated after providing so many devastating arguments against the teleological argument. Thus far I go along the
same path as Dawkins.

But what happens next? Darwinism may open the way to atheism, but
does it necessitate atheism? Does it necessitate a rejection of Christianity?
Rejecting Christianity is a rather weaker option than accepting atheism (one
could still accept deism for instance); but perhaps an argument for something along these lines influenced Darwin, particularly in his debate with his
great American follower Asa Gray. Darwin's point in this debate seems to be
that, although the world is indeed design-like, the mechanism of natural
selection precludes any kind of God except in a very distant sort of way:
eighteenth-century deism rather than nineteenth-century Anglo-Catholicism
(Moore 1979: 274). Darwin's argument bears on the unlikelihood that the

Christian God would have been quite as indifferent to organic need as selection supposes at this point.

Interestingly Dawkins stands, with respect to this line of argument, very much in the tradition of Sir Ronald Fisher (1950), rather downplaying the whole significance of variation. At the least, he downplays the significance of the randomness of variation – and thus coincidentally removes any bias toward deism and away from Christianity. Fisher's point was that in some sense variation is so common and so small that it swamps out or eliminates the effects of its randomness – the randomness becomes unimportant. A similar sort of argument is endorsed by Dawkins, particularly in a brilliant chapter of *The Blind Watchmaker* in which he shows how computer programs can, very rapidly indeed, generate order from randomness. Dawkins reduces the randomness of mutation to a mere technical detail. Randomness is not something with profound implications, and certainly not something with profound theological implications. It is simply the raw material on which evolution builds: the fact that it is random is really quite irrelevant given the swamping nature of the selective process. Of course, the randomness is important for Dawkins in other respects: it does mean that evolution is not directed toward some pre-ordained plan which we know in advance will be fulfilled. Selection is opportunistic. Not only are there no immediate good mutations but the very standard of "goodness" is relativized – which at the least is going to call for some deep thinking by the Christian who surely supposes that the emergence of humankind was not pure chance, but in some sense intended by God.

However, Dawkins differs strongly from Stephen Jay Gould over the progressiveness of evolution. Against Gould's claim (in such places as his book *Wonderful Life*), Dawkins argues that the evolution of intelligence is not just a matter of chance. Dawkins promotes something which he calls the "evolution of evolvability" – something which is a kind of almost preordained upwards stepping of the evolutionary process (Dawkins 1988; Dawkins and Krebs 1979). Whilst I am sure that Dawkins does not think that humans had to evolve exactly as they are, he certainly thinks that our evolution is more than just chance – in the sense that, given natural selection, anything could have happened. He invokes the idea of an "arms race" where lines of organisms compete against each other, improving adaptations: the prey gets faster, the predator gets faster. Dawkins (1986, 1997a) thinks that intelligence is something which virtually had to emerge out of this process, as electronic devices (like computers) have emerged from and now dominate human arms races. Thus, I am not at all sure that Dawkins thinks that the evolutionary process is so very non-directed. In this respect, a Christian might well claim Dawkins as an ally.[5]

What about the next string to Dawkins's bow: the argument from evil? Dawkins thinks that the problem of evil is incompatible with a good God. More than this, Dawkins clearly thinks that natural selection – relying as it does on a struggle for existence and the production of adaptations which will aid organisms in the struggle for existence – leads to an intensification of the problem. The evil was there and identified before Darwin set to work, but natural selection not only draws attention to it, but essentially suggests that evil and pain are bound to come through the selective process. This is something that one would expect from the most basic workings of nature, rather than existing (as it were) tacked on. Dawkins does not get into detailed theological argument, but his case seems to be that, in the world that we have, evil is neither contingent – perhaps a result of human action – nor something readily eliminable. It is as much a part of the essence of the organic existence as it is possible for anything to be. It is something incompatible with a good God, nor (Dawkins does not bring this out explicitly) does it seem that traditional counters will make it acceptable. It is stretching credulity to suppose with Augustine that evil is merely a privation, a lack of good. Selection positively produces evil – the parasite who exists only through the destruction of its host, for instance.[6] Nature is, as he says, "pitiless."

I should say that I am not entirely happy with all of Dawkins's metaphors at this point. Emotional dislike of religion rather than reasoned discussion is driving this argument. Heinrich Himmler was pitiless toward the Jews. It is inappropriate to speak of the laws of nature as being likewise pitiless. Things which are pitiless are things which ought to show pity, but which do not. That is why Himmler was such an evil person. The laws of nature are not things which should in any sense show pity. To speak of them as pitiless is a rhetorical trick: straining the metaphor in directions favorable to Dawkins's atheistic conclusion. Thus, Dawkins's metaphors make me somewhat tense, although we all know full well why he has chosen them (Ruse 1999).

But let us go to the main argument about the problem of evil. This is indeed a significant problem for the Christian believer. Moreover, Dawkins is surely right in thinking that natural selection does draw attention to the existence of evil – an evil which is particularly acute for the Christian believer, because it involves precisely the kinds of pain and discomfort which cannot be explained away by the traditional free-will defense. Things like the cheetah eating the gazelle are not things where we think naturally of free will. At least one can say this: even if the cheetah has the freedom not to chase the gazelle, an omnipotent, omniscient God might have decreed things otherwise, since a vegetarian cheetah would last a very short time in the wild. Its eating apparatus is not suited for a vegetable diet, neither is its stomach. Gazelles conversely would probably suffer through over-breeding and so forth. So I do think that we have a serious problem here for the Christian believer. However, this is a problem that the Christian believer surely had all along and has had to wrestle with quite irrespective of Darwinism. Although to a certain (significant) extent Darwinism intensifies the problem and makes it more acute, it is certainly not something that one had to be a Darwinian to appreciate. If Darwinism points to atheism, it is not a new signpost.

It is worth remembering that Fabre, on whom both Darwin and Dawkins are relying at this point, did not feel that his discoveries were such as to plunge him into a Darwin-inspired atheism. It is true that his position might not have been ultimately tenable, but it does make one wary of assuming that all discussion is now closed. Perhaps, then, the Christian does have an answer to this kind of pain and discomfort. Perhaps the Christian does not. But this is more the Christian's problem than the Darwinian's special insight. My own feeling is that if such an answer can be given, it will be one which comes in some way because of the pain and discomfort – the pitiless-ness of nature in Dawkins's language – rather than despite it. Darwinism will thus become part of the solution rather than despite it. My point is Kierkegaard's, that without the human experience of pain and trial here on earth, faith would become altogether too easy and meaningless – the separa-tion and alienation from God at the heart of Christianity would be a sham. The crucifixion would become unnecessary. Frankly, I am not at all sure that this line of argument can be pursued to its end. If it can be done, then as I said, not only does the Darwinian position fail to negate Christianity, it becomes part of the solution. But my main point here is that, whether or not the Christian has an answer, Dawkins has not shown that Darwinism as such is a crucial motivating force toward atheism. Someone who is an atheist already on the grounds of evil would find Darwinism a comforting support; but if one rejects atheism despite evil, I do not see that Dawkins has provided any argument for suggesting that one ought to change one's mind.

Daniel Dennett

It would be unfair to describe the philosopher Daniel Dennett simply as cleaning up loose ends from Dawkins's position, but it is certainly fair to say that Dennett follows very much from Dawkins. He accepts the science in the way that Dawkins does and he accepts the same kinds of philosophical inferences. In his most recent book on the subject, *Darwin's Dangerous Idea*, he reiterates Dawkins's point about Hume being caught in the fix of knowing the fallacious nature of the argument from design but not having a satisfactory alternative. Where Dennett goes beyond Dawkins is in criti-cizing anyone who wants to argue that, although Darwinism may be correct, one can still accept design. He wants to counter the person who says that this design can be accepted on faith or some such other non-scientific grounds (Dennett 1995: 154–155).

At the risk of sounding like a broken record which simply repeats itself, let me say again what I have said above: although I think that Dennett certainly puts his finger on significant problems for the Christian theist, I fail to see in any way that these are problems which are brought on specifi-cally by Darwinism. I would grant that there are problems with – certainly questions which need answering about – Christian faith. I would grant

Dennett that, all too frequently, faith is simply a form of irrational commit-ment to believe something because you want to believe, despite all the evidence to the contrary. Faith in the promise of eternal life is often much akin to faith that one's lottery ticket will win the grand prize: in both cases one wants something very much and so one has willed oneself to believe that one will get it.

But suppose one takes the position that there is more to faith than – as at the end of the last paragraph – autobiographical references to the hopes of the infant Ruse. Suppose one argues that existence poses questions which are simply unanswered by science – Why are we here? Why do good people suffer? and so forth – and that the fact that science fails to answer them is no good reason for saying that they are unanswerable or that we should not seek answers. Suppose that one says that it is appropriate to turn to some-thing outside of science and that faith (however construed) can start to bridge the gap. Even if one disagrees with this move, I do not see that Dennett has shown that Darwinism strengthens the disagreement. Dennett seems to think that Darwinism implies materialism, but he has given no grounds for believing that this is so. Apart from the fact that, in this day and age of electrons and so forth, materialism is a bit of an old-fashioned philos-ophy anyway, it certainly seems to me possible for someone to be an ardent Darwinian evolutionist and yet argue that the mind is not something which can be reduced to material entities. The late Karl Popper (1976), for one, was both a Darwinian and a mind–body dualist.

As is well known, it is true that Dennett himself is a materialist with respect to minds, thinking that minds are somewhat akin to the software used by the hardware of our brains. But others, most particularly the Berkeley philosopher John Searle (1997), would dispute this strongly. And I too see no reason at all why one should not be a Darwinian evolutionist and think that in some sense minds involve the non-material in some sort of way. Not a mysterious non-material substance akin to a life-force or vitalistic entelechy or *élan vital*, and certainly not necessarily a supernatural non-material substance, but more than just material physical objects. And analogously – if the non-material is not ruled out by fiat – it seems quite possible for someone to argue that they are a Darwinian but that they think that this proves design, rather than detracts from it. I am certainly not saying that you should go this route. What I am saying is that, if you want to go this route, there is nothing in Dennett (other than blunt assertions) to stop you. Indeed, there is nothing in Dennett to stop you from arguing that Darwinism confirms and strengthens the inference rather than the other way around.

Conclusion

My conclusion is clear. For all that several of today's most prominent Darwinians have argued that their scientific theory throws significant doubt on Christian belief, even to the point of inviting one to embrace atheism, ultimately none of the arguments is definitive. The Christian ought to take Darwinism seriously and this may well demand modification of some of the beliefs that the Christian held before the *Origin of Species* was published. But, ultimately, nothing in Darwinism absolutely forbids a belief in Christianity. Perhaps indeed there are things in Darwinism which the Christian might find comforting.

This being said, let me conclude with one final thought – something made most clearly by the great twentieth-century evolutionist J.B.S. Haldane (1932) – one on which anyone, Darwinian or Christian, ought to reflect most carefully. Haldane pointed out that there is absolutely no reason to think that mid-range primates – such as we humans – have evolved the ability to delve into the ultimate mysteries of reality. He argued that not only may the real world be stranger that we think it is, it may even be stranger than we could possibly think it is. Haldane concluded that accepting Darwinism ought to instill in one a sense of modesty, not only about our abilities, but about the range over which our abilities can reach and grasp.

This reflection by Haldane does not in itself mean that Christianity at once becomes a reasonable position. Indeed, I think it points one toward a modest skepticism. But this is a skepticism (perhaps not so very modest!) which ought to be extended also to atheistic pronouncements by people like Dawkins and Dennett. Perhaps there is nothing beyond our ken, and perhaps the atheists are right. But as a Darwinian I think that one ought to be very careful in making such pronouncements. I have yet to be convinced that Darwinism and atheism is a marriage made in heaven.

Notes

1   As it happens, I have written extensively on these matters elsewhere (Ruse 1982, 1988). So my silence does not imply a lack of interest.
2   There is more to evolution than Darwinism and there is more to religion than Christianity, but it is the supposed clash between these two belief systems which has been the focus of recent attention and on which I concentrate in this essay.
3   Technically: the truth of evolution is a necessary condition for the soundness of Wilson's argument; it is not necessarily a sufficient condition.
4   Although the ubiquity law was taken as proof of God's distance in the hands of the deists, it was taken as proof of God's constant presence in his creation as rejigged in the hands of Anglo-Catholics like Aubrey Moore.
5   For more on this whole question of progress in biology, see Ruse (1993, 1996) and McShea (1992).
6   Note that Dawkins's desperate desire to get to atheism takes him right past another Christian heresy, that of the Manichaeans. He does not want to argue that natural selection points us toward the existence of an evil God. Rather, he argues that there is nothing at all.

References

Dawkins, Richard (1986) *The Blind Watchmaker*, New York: Norton.

—— (1988) "The evolution of evolvability," in *Artificial Life*, edited by C.G. Langton, Redwood City, Calif.: Addison-Wesley.

—— (1995a) "Richard Dawkins: A Survival Machine," in *The Third Culture*, edited by J. Brockman, New York: Simon & Schuster.

—— (1995b) *A River out of Eden: A Darwinian View of Life*, New York: Basic Books.

—— (1997a) "Obscurantism to the rescue," *Quarterly Review of Biology* 72: 397–399.

—— (1997b) "Religion is a virus," *Mother Jones* November.

Dawkins, R. and J.R. Krebs (1979) "Arms races between and within species," *Proceedings of the Royal Society of London, Series B: Biological Sciences* 205: 489–511.

Dennett, Daniel C. (1995) *Darwin's Dangerous Idea*, New York: Simon & Schuster.

Fisher, Ronald A. (1950) *Creative Aspects of Natural Law*, The Eddington Memorial Lecture, Cambridge: Cambridge University Press.

Gould, S.J. (1989) *Wonderful Life: The Burgess Shale and the Nature of History*, New York: W.W. Norton Co.

Gould, S.J. and R.C. Lewontin (1979) "The spandrels of San Marco and the Panglossian paradigm: a critique of the adaptationist program," *Proceedings of the Royal Society of London, Series B: Biological Sciences* 205: 581–598.

Haldane, J.B.S. (1932) *The Inequality of Man and Other Essays*, London: Chatto & Windus.

Hume, David (1947) *Dialogues Concerning Natural Religion*, edited by N.K. Smith, Indianapolis: Bobbs-Merrill Co.

McShea, D.W. (1992) "Complexity and evolution: what everybody knows," *Biology and Philosophy* 6.3: 303–325.

Moore, J. (1979) *The Post-Darwinian Controversies*, Cambridge: Cambridge University Press.

Popper, K.R. (1976) *Unended Quest: An Intellectual Autobiography*, LaSalle, Ill.: Open Court.

Ruse, Michael (1982) *Darwinism Defended: A Guide to Evolutionary Controversies*, Reading, Mass.: Benjamin/Cummings Pub. Co.

—— (1986) *Taking Darwin Seriously: A Naturalistic Approach to Philosophy*, Oxford: Blackwell.

—— (ed.) (1988) *But is it Science? The Philosophical Question in the Creation/Evolution Controversy*, Buffalo, N.Y.: Prometheus.

—— (1989) *The Darwinian Paradigm: Essays on its History, Philosophy and Religious Implications*, London: Routledge.

—— (1993) "Evolution and progress," *Trends in Ecology and Evolution* 8.2: 55–59.

—— (1995) *Evolutionary Naturalism: Selected Essays*, London: Routledge.

—— (1996) *Monad to Man: The Concept of Progress in Evolutionary Biology*, Cambridge: Harvard University Press.

—— (2000) *Can a Darwinian be a Christian? The Relationship Between Science and Religion*, Cambridge: Cambridge University Press.

Searle, John (1997) *The Mystery of Consciousness*, New York: New York Review of Books.

Tort, P. (1996) "Jean-Henri Fabre, 1823–1915," in *Dictionnaire du Darwinisme et de L'Évolution*, edited by P. Tort, Paris: Presse Universitaire de France.

Wilson, E.O. (1978) *On Human Nature*, Cambridge: Cambridge University Press.

11 Geoffrey F. Chew

Geoffrey F. Chew is a retired Professor of Theoretical Physics at the University of California–Berkeley who is continuing research in quantum cosmology at the Lawrence Berkeley National Laboratory. At UCB he served as Chair of the Physics Department and later as Dean of Physical Sciences. He is a member of the National Academy of Sciences and of the American Academy of Arts and Sciences. For his work on the bootstrap theory of nuclear particles he received the Hughes Prize of the American Physical Society and the Lawrence Prize. He is author of two books as well as many scientific papers.

Interview by Philip Clayton

PC: Do you identify with any particular religious tradition?

GC: I am Roman Catholic.

PC: Did you grow up in the church, or was that a later influence?

GC: No, it was later – the result of my second marriage, as a matter of fact. My mother was brought up in the Church of England, and my father was an agnostic. I did get a kind of Christian education. However, my joining the Roman Catholic Church was the result of my marriage about twenty-five years ago to a French woman, Denyse Mettel, who was deeply committed to the church. She impressed me with her convictions, so I decided to join.

PC: Did questions of integration between religion and science arise for you only later in life, or were they present for you in childhood and as you began your scientific training?

GC: They were present earlier, though not specific. The religious exposure from my mother, although of a traditional type, already started me doubting in many directions about the posture of institutionalized religion. I went through the phase of believing that religion was of no use because it conflicted with scientific knowledge. But at some point I became aware of the limitations of science. It's a little hard to say when, but as I got into my own research I became aware of questions inaccessible to the scientific approach.

PC: What were the questions you began to recognize couldn't be answered by science?

GC: I was a student of Enrico Fermi at the University of Chicago. Fermi was a supreme pragmatist. He was not interested in anything that could not be tested by the experimental method. He disdained the philosophical problems surrounding quantum mechanics. For a long time, as a result of this exposure to Fermi, I thought such questions could not be interesting, that they were not something one should spend time thinking about.

PC: But later you became dissatisfied with that approach?

GC: Curiously, Fermi's attitude came into conflict with my actual research. It was only when I started trying to find answers to questions I thought were scientific in nature that I began to appreciate the methods of science were fundamentally limited. The first such issue I became involved with was the concept of elementary particles, which Fermi had accepted very comfortably. I was one of the early people to come to believe that the 'elementary particle' concept was fundamentally flawed and could not be a basis for indefinitely increasing our understanding of nature, although it was only after Fermi died in 1953 that I was able to begin to develop such feelings. A curious aspect of the approach I have since developed, which may not be rational but has allowed me to keep moving forward, is that I have a dream of formalizing some of these questions that go beyond conventional science. I want to get outside the standard scientific notion that certain phenomena can be isolated.

PC: Is there any connection between your scientific approach and a religious response to the universe?

GC: Connections with Eastern religious traditions are more immediate than with Western traditions. I was first made aware of this by my son, who had a course in Buddhism at Princeton in the early 1970s. In the Buddhist approach you manage to avoid starting with objects in your thinking. It is holistic rather than objective. Objects derive significance from the totality.

PC: Is the parallel between your work and the Buddhist approach significant for you?

GC: I'm not sure. This line of thought encourages me when I think about the mysterious notion of "consciousness." I favor the view that consciousness is not localized, that it is a concept going beyond life and has a broader significance. I was astonished to discover recently that this way of thinking about consciousness is familiar to certain Buddhist traditions. One of the reasons the usual scientific approach to consciousness gets blocked is that science normally defines itself in terms of isolated objects. Once you've done that, you have closed the possibility of progress with consciousness.

PC: Does this tendency in science toward isolating phenomena create unacceptable obstructions by turning away from the whole and studying only parts?

GC: Yes, but I think the word 'unacceptable' is a little too strong. The success of the standard scientific approach is striking. One of the major mysteries that needs clarification is why it works so well. Why is it so efficacious to think of the universe as made up of separate objects, isolatable systems? There has to be some explanation.

PC: Do you have any hypotheses?

GC: The only guesses I have come up with reflect the idea that a very large or a very small parameter can create the *appearance* of absolute truth. For example, the fact that Planck's constant is very small has allowed humanity until recently to think of objectivity as an absolutely valid basis for looking at the universe. This has turned out to be only an approximation. Objective reality works only because the Planck constant is so small. The fact that the velocity of light is large allows us to contemplate many aspects of our surroundings without worrying about funny notions of Lorentz contraction and time dilation. Physicists are familiar with large or small parameters that create the illusion of truth for certain ideas that are only approximate.

I deeply believe that absolute truth is unobtainable, that we are never going to have a basis for any absolutely true statement about the universe. There will only be approximate statements. This is perhaps the most religious statement I can come up with. I firmly believe that there are no absolutely true statements that human beings can ever hope to make about the universe. Truth is inherently unattainable.

PC: How do you relate the pursuit of truth in science to the religious pursuit of truth?

GC: I'm not quite sure what religion means. It seems to me that religion has mostly to do with human questions, while science asks a broader set of questions. However, if by religion, and some associated notion of God, you mean the entire universe, then I don't see any difference between religion and the kind of science I am interested in.

PC: The monotheistic traditions affirm beliefs about the purpose and destiny of the universe. How do these notions fit in with perspectives shaped by contemporary physics?

GC: I don't really know what destiny or purpose means. I can imagine that as the universe expands, qualitatively new domains of reality emerge. That appeals to me, and I don't see any conflict between that and any of the notions that I'm following. It connects with the idea that truth is never absolute. There are always different levels of phenomena, and new types of phenomena may emerge in the future that we can't even imagine now.

PC: The monotheistic traditions each focus on human beings as persons who bear a special capacity for relationship with God. How does this fit with contemporary science?

GC: I have a lot of trouble with the idea of the person as a separated entity; I tend to see the notion of an individual as some kind of an approximation. The boundaries in space and time are not sharp. I start this sentence and end it as a different individual, physically different. The mystery is, why do we get as much benefit as we do from thinking of ourselves as an individual? It has something to do with the fact that the changes are relatively slow. There is a meaningful scale in which you can say I'm only slightly different in the next moment than I was previously. The difference is not enough for me to worry too much about.

PC: So the fact that we would be unlikely to view ourselves as enduring entities or substances if the rate of life increased by an order of magnitude or two, makes you think our notion of substance, whether human or divine, is fundamentally flawed?

GC: Yes.

PC: Is there any difference between religion and science in this broadest sense?

GC: I think about religion more in the sense of Buddhism than I do in the sense of the Western religions. It comes down to the fact that the importance of the individual is paramount in Christianity, Islam and Judaism. It is much less important for Buddhism.

PC: Might I ask if you practice your own Roman Catholicism in a manner similar to your science?

GC: I suppose so. One of the things that appeals to me about Catholicism is that it isn't well defined. It allows a tremendous breadth of interpretation.

PC: Is there anything else you would like to add?

GC: It seems to me that "religion" means something different from "physics" as it is traditionally defined, based on the reproducible experiment, the isolatable object. Because I'm deeply convinced of the limitation for this meaning of physics, convinced there is no "theory of everything" as some physicists expect, then I tend to say, "OK, I must be religious."

Geoffrey F. Chew

An historical reality that includes Big Bang, free will and elementary particles

Introduction

This essay forecasts the development of a "natural science" based not on objective reality but rather on what Whitehead (1929) called an "historical reality." Such a science, embracing future as well as past, would come to grips with the hitherto-inaccessible phenomenon of *free will*. In the foreseen scientifically meaningful historical reality, there is a parallel with what David Bohm (1980) has called "implicate order." The envisaged broader reality would combine local Whiteheadian "process" with quantum principles and the global evolution of the universe.

Objective reality, currently epitomized by representation of the universe through the elementary particles of grand unification theory (GUT), will be recognized *within* global historical reality as a localized causal *component* of universe history – the only portion of reality that science has so far been able to define. I expect quantum cosmology to elaborate Whitehead's "process" so as to allow objective reality (e.g., causally interrelated elementary particles) to be recognized not *a priori* but as a consequence of a universe evolution that includes the impact of free will.[1]

Whitehead is identified as a philosopher, not as a scientist. Why do I expect his approach eventually to reshape natural science? A superficial consideration is that Whitehead's "process" invoked a concept whose precise definition requires mathematical language of a type developed by physicists, not by philosophers. A more substantial consideration is that, in the face of growing human appreciation of evolution, physics cannot indefinitely sustain a "purely objective" stance that disregards the history of the universe. "Process" puts history up front while managing to recognize objective reality as an important feature of an "adult" universe, even though not the only feature and a feature absent when the universe began.

Whitehead's "process" involves a chain of *events* individually localized both in space and in time. At a Whiteheadian event something "happens" although not to an object. The notion of object is subordinated to the notion of "happening." "Objects" correspond to certain very special *patterns* of events – patterns that may or may not develop.

The foregoing may be restated in familiar physical language by saying that Whitehead subordinates energy to *impulse*.[2] For Whitehead, localized energy is not *a priori*; the "happenings" that build our universe are *impulses* localized in spacetime. When a pattern of impulses exhibits a certain persistently repetitive regularity, the *frequency* of repetition provides (via Planck's rule) meaning for energy. But not all impulse patterns need be repetitive: historical reality has both objective and non-objective content.

Because free will, whatever the precise meaning of this term, impacts history, historical reality includes free will. The "commonsense" belief that free will is located *outside* objective reality might be an illusion; perhaps free will is no more than a manifestation of causally behaving elementary particles; perhaps historical reality adds nothing to objective reality. I do *not* believe such to be the case.

One influential consideration stems from the cosmological idea of the "Big Bang." Studies of this notion by many different workers using many different approaches concur in concluding that in the extremely early universe *all* "ordinary" physical concepts, including both particle and field, break down. The meaning of local causality breaks down. Historical reality, if it is to embrace an early as well as a late universe, must include non-objective components.

A second set of considerations stems from my own efforts to represent historical reality through a chain of events consistent with the Big Bang scenario, each event locating an impulse. A promising event-pattern has been found for representation of elementary particles but *all* events in a Whiteheadian chain cannot congregate in such patterns. I shall expand below on this assertion.

I should dearly like in this essay to report discovery of an event-pattern associable with free will. This I cannot do. But the richness of encountered patterns is to me persuasive of there being far more than objective content in historical reality.

Let me now consider more precisely the meaning I attach to "objective reality," making a link to the traditional meaning of natural science.

Objective reality

The term "objective reality" as used in this essay refers to those aspects of our universe characterizable as "material" in the extremely general sense introduced by Einstein. In Einstein's sense, "matter" is synonymous with *localized energy*, so light and gravity as well as atoms are forms of matter. Both fields and strings are "material" in the sense of this essay.

Ever since the astounding successes of Isaac Newton three hundred years ago, physics (originally called "natural philosophy") has been based on energy localized in space and time. Twentieth-century science was forced by experimental observations of matter at small scales to introduce a set of principles called "quantum mechanics" – principles that entail non-causal uncertainty aspects which continue today, after decades of study, to baffle the wisest humans.[3] Often these "weird" aspects are called "non-objective" and I agree that they should be so described. But quantum mechanics has failed to dislodge localized energy – i.e., "matter" – as the underpinning of natural philosophy. What this essay calls "objective reality" includes not only gravity, light, electrons, and any other forms of "matter" (such as strings), but the Hilbert-space representation thereof as described in the most advanced physics graduate courses. The term, "Hilbert space," characterizes the mathematical language with which physicists state quantum principles. Maintaining Newton's materialistic worldview, quantum physicists have heretofore supposed that all Hilbert-space labels refer to "matter." (To what else might these labels conceivably refer? The reader will by now not be surprised to hear that Hilbert-space labels may refer to history.)

The achievements of a science based on matter have been so impressive as to raise hopes for an eventual "materialistic theory of everything." Such a theory would embrace many important issues. How is it conceivable, for example, that such phenomena as consciousness and attendant free will might be manifestations of localized energy?

A persuasive consideration is that free will affects the behavior of matter: I decide to raise my arm and matter responds. Given such linkage between matter and free will, is it not reasonable to seek understanding of *all* reality through energy? Success for a materialistic theory of everything would mean absence from universe history of "non-objective" content.

I have by now repeatedly indicated my failure to envisage complete success for materialism. In particular, I believe free will to be a component, albeit a non-objective one, of historical reality.

A model of historical reality

My expectations for the future of science have been influenced by a cosmological model I have been pursuing over the past decade in a search for the quantum mechanical meaning of space and time.[4] Several years ago I began to connect the model with Whitehead's idea of a chain of history – a global universe history comprising a huge collection of discrete, ordered events. I found in the model *explicit* realization of a key Whiteheadian supposition: that "objects" – i.e., "matter" – correspond to regular localized *repeating* patterns of large numbers of events. The model locates an impulse at each event; spacing between successive events is on the scale of Planck time – roughly 10^{-43} seconds. The identity of an "object," such as an electron or the reader of this essay, resides in the detailed structure of its repeating event-pattern.

A quantum mechanical consideration, probably grasped by Whitehead although seemingly ignored in his writings, is that *energy* associates with *frequency* of pattern repetition. Thus energy derives from *regular* collections of impulses.

But not all impulses lead to energy. I found in the model that many events are unable to build patterns exhibiting the causal regularity characteristic of matter – i.e., characteristic of energy. This model of historical reality, in other words, has explicit non-objective content. In fact, because a single chain of Whiteheadian events constitutes the history of the universe, with every event in principle influencing and being influenced by every other event, it is at first sight difficult to imagine *any* meaning for objectivity in historical reality. How might there be particles obeying local Einsteinian causality within a global Whiteheadian chain of history?

The mechanism provided by the model depends on screened magnetism[5] in combination with a meandering of the history-chain back and forth in local time. Within spacetime regions that are large on a Planck scale while tiny on a human scale, a tightly woven, closed two-dimensional fabric of causally related events is possible. Stability of this "material fabric" is electrodynamic. Different fabric patterns correspond to different forms of matter. Elementary particles in the GUT sense correspond to *tubular* event-patterns, with the radius of a tube being roughly a hundred times larger than the Planck length (10^{-35} meters). This factor relates to the fine structure constant.

Events at which there is reversal of time direction cannot be part of material fabric. Although certain events of such temporally "indecisive" character associate with the birth and death of elementary particles,[6] there remain in the model additional events with "a-causal" relation to adjacent events in the chain of history. Interpretation needs to be found for "loose indecisive strands" of history that lie outside material fabric even though comprising events which, individually, are similar to those building objective reality. Such loose strands influence the objective component of history, although not "causally" – i.e., not in the manner, according to the laws of objective physics, by which one piece of matter influences another.

Freedom from the causal constraints of tightly woven material fabric suggests for "loose" event-strands an historical role that is unstable and unpredictable. This role cannot be described by a causal materialistic science that recognizes only the objective component of history. Might loose strands underlie what we call "free will"? I have been unable to find a persuasive argument to the contrary.

How can "scientific" study of "loose event-strands" even be imagined? How can questions, to say nothing of answers, be formulated? I believe the response to this query lies in mathematics and the requirements of consistency. In my model, appreciation of the very special two-dimensional event-fabric interpretable as matter comes through mathematics. Ordinary language, with all its nouns, is so prejudiced by objective reality as to be unsuited for asking and answering questions about historical reality, but mathematics provides such a language.

One immediately identifiable limited goal, closely connected to objective reality, is a mathematical meaning for *measurement* within historical reality.

Any measurement is a part of history, and human ingenuity may allow isolation of the essential mathematical characteristics of a Whiteheadian event-pattern that qualifies as a measurement on "one piece of matter" carried out by another. An associated, but less challenging to define, event-pattern would be "record of measurement" or "memory." Artificial intelligence theory, already under development, promises here to be of assistance. Historical meaning of measurement will interlock with but probably not be contained *within* material fabric. *Record* of measurement, on the other hand, promises to reside in this fabric. Memory, after all, is a component of an object's identity.

Any individual measurement is a portion of history and its event-pattern links the measurement to "surrounding history." *Why* did the measurement occur? Such a question parallels, Why did elementary particles emerge from an early universe where particles had no meaning? Contemporary physics-cosmology is attacking the latter question with ideas such as that of inflation. Why should science not contemplate the "reason" behind measurement? You may say, "Because the reason involves free will and science is helpless to deal with free will." But I say free will is part of the history in which matter emerged from non-matter.

The a-causal aspect of "loose events" does not mean such events are "above the law," following no rules. They are part of a single Whiteheadian chain of history and, at least in the model I have been studying, enjoy electromagnetic connections to other events. The meandering of loose event-strands of history, even though a-causal and *locally* unpredictable, is subject to *global* constraints – free will is cosmologically controlled.

A global character need not put free-will investigation beyond human reach. Cosmology has proved to be a fruitful area for human enterprise, even though some of my most respected physicist colleagues claim inability to "think about the entire universe."

Recognition of measurement as simply part of Whiteheadian history would eliminate the collapse of the wave-function as a separate puzzle. I see one mystery, not two. Connection between free will and quantum mechanical "collapse" has often been conjectured. What about free will's cosmological aspect, its connection to Big Bang? Like all models, the one I have been studying allows arbitrariness in global "initial conditions." Once it is acknowledged that mysterious global conditions influence loose event-strands throughout history, one appreciates that none of the speculations in this essay diminishes God's role in the universe. But agreeing that God's role may be manifested through free will does not mean to me that effort is futile to improve understanding of Whiteheadian events located outside the fabric of material reality.

My association of objective reality with localized energy might seem not to guarantee the ordinary meaning of "objectivity," which implies a reality *independent of observer*. When, in a criticism of Copenhagen quantum mechanics, Einstein insisted that "the moon is really there," he expressed his desire for a universe whose meaning transcends observers and measurement. Historical reality, by recognizing observers and measurements to be merely aspects of history, accords with Einstein's wish. I must nevertheless now report a bizarre feature of my historical reality model – a feature troubling to colleagues. The model's accommodation of Big Bang renders history dependent on "standpoint" – the meaning of standpoint being "here and now." I cannot guarantee that any conceivable model of an expanding universe historical reality will exhibit standpoint dependence of history, but this one does.

My colleagues have been placated by learning that the *objective* component of history occurring *close* to the standpoint is independent of standpoint. "Close" means on a Hubble scale, so all standpoints occupied by humans, assuming a duration of our species small compared to the age of the universe, are "objectively equivalent." Humanity shares a common objective reality, at least according to the model I have been discussing here.

On the other hand, the model's *full* historical reality appears to be standpoint sensitive. Loose event-strands have a tendency to congregate spatially near the standpoint. Conversely, one might say that (total) history becomes "more objective" as the distance of events from standpoint increases. Such a statement, while sounding philosophical rather than scientific, enjoys mathematical representation in the model.

Conclusion

Two decades ago David Bohm (1980) wrote an intriguing little book distinguishing what he called "implicate order" from "explicate order." Although Bohm ignored the evolution of the universe, I like to associate what I have called "the objective reality of localized energy" with Bohm's explicate order. Explicate order would comprise that very special component of historical reality that is bound up in tightly woven material fabric located close to the standpoint. Implicate order would comprise *complete* historical reality – a meandering chain of events only a portion of which lies within objective fabric. The remainder, inaccessible to objective science and embracing free will, is nevertheless essential to the whole.

It may be useful to appreciate that, from *our* standpoint (here and now), the total number of Whiteheadian events in our history is huge but not infinite, depending (unsurprisingly) on the age of the universe. Because the total number of events increases with the age of the universe, it is plausible that qualitatively new patterns of events will develop indefinitely. Remember that the tightly woven fabric of objectivity could not exist when the universe was very young. Conversely, when the universe is much older than it is today, event-patterns may develop with a significance that goes beyond that of matter and free will.

Presently, my efforts are focusing not on free will but on details of the material fabric interpretable as light. My longstanding identity as an objective physicist compels me to give higher priority to light than to free will. After light the fabric of gravity must be understood. After gravity will come the event-structure of particles such as electrons. But exertions in my identity as a physicist will not cause me to forget about free will. Understanding *anything* about an event-pattern associable with free will would for me eclipse in satisfaction any imaginable "physical" illumination generated by contemplation of historical reality.

Notes

1   The importance of global evolution has led to my preferring the adjective "historical" to Whitehead's term, "process." The term "process" carries a local connotation.
2   The psychological meaning for the word "impulse" – a precursor to exercise of free will – is curiously suggestive of free will's inclusion within historical reality.
3   The non-causal implications prevented Einstein from ever accepting the "completeness" of quantum mechanics.
4   Those interested in the model's technical details can see my "Standpoint Cosmology" (1995).
5   Because equal quantities of positive and negative magnetic charge locate inside an elementary-particle history pattern, there are no magnetic-charge effects outside this pattern.
6   Particle birth or death corresponds to the more usual notion of "event" – *not* the Whiteheadian notion. The time scale characterizing particle birth and death lies above the Planck time scale even though far below the human scale.

References

Bohm, David (1980) *Wholeness and the Implicate Order*, London: Routledge & Kegan Paul.
Chew, Geoffrey (1995) "Standpoint Cosmology," *Foundations of Physics* 25 no. 9: 1283–1333.
Whitehead, Alfred North (1929) *Process and Reality*, New York: Macmillan.

12 Michael A. Arbib

Michael A. Arbib is not only a neuroscientist and computer scientist but also that strange creature, an atheist theologian. In 1983, he and Mary Hesse gave the Gifford Lectures in Natural Theology at Edinburgh University on the theme "The Construction of Reality" and, more recently, he acted as the neuroscience organizer and editor for the series of workshops that resulted in *Neuroscience and the Person: Scientific Perspectives on Divine Action* (Vatican Observatory and CTNS, 1999). He is currently University Professor and the Fletcher Jones Professor of Computer Science at the University of Southern California (USC), as well as a Professor of Biological Sciences, Biomedical Engineering, Electrical Engineering, Neuroscience and Psychology. He directs the USC Brain Project, an interdisciplinary project integrating studies of brain mechanisms of learning with research on neuroinformatics. Arbib is the author or editor of more than thirty books including the edited volume *The Handbook of Brain Theory and Neural Networks* (MIT Press, 1995), and *Neural Organization: Structure, Function, and Dynamics* (MIT Press, 1998), co-authored with Péter Érdi and the late John Szentágothai.

Interview by Gordy Slack

GS: Would you say a few words about your own religious background?

MA: I come from a Jewish family. I was born in 1940 in wartime England while my father was in the army. About a year later, he went to serve in the North African desert and was a prisoner of war throughout the rest of World War II. Our family was not particularly religious; I don't remember going to synagogue during that period in England. On the other hand, my mother was working with Jewish organizations.

When I was seven we moved to New Zealand, where we lived in a small town about twenty miles outside the capital of Wellington. There was no local synagogue. When I was nine we moved to Sydney, Australia, where my parents became involved in the local synagogue and also in B'nai Brith, the Jewish organization. Ours was more a "high holy day" type of Judaism than a weekly ceremony in the home. Because I went to a Presbyterian boys' school, I had a lot of exposure to Christianity. In fact, I observed services in a number of churches, and from that I learned quite a lot about both the Old and the New Testament, but was sufficiently Jewish not to believe all the stories in the New Testament. That may have fermented a certain skepticism, since if I wasn't to believe part of the Bible, why should I believe the rest?

By the time I was fourteen or so, I was an agnostic. Perhaps the turning point was a divinity class I took at this Presbyterian school. I decided to study for the exam and finished at the top of the class. Applying a strange teenage logic, I wondered what sort of God would let a Jewish boy be the top of a Presbyterian class? This made me less devout than I might have been.

I have been interested in religion, have in fact thought a lot about the basic questions of religion, have gained an understanding of different people's faith, but have been agnostic for a long time. By now I would declare myself an atheist, but an atheist who is very interested in the study of religion. I would also say I'm a Jewish atheist. Over the years as I came to better understand both Jewish history and what had happened during the Holocaust, I built a strong cultural identification with what the Jewish people have been through over the millennia. I have no desire to renounce that background.

As a scientist, my faith is that I can make continuous progress in finding an explanatory system which will make sense of a whole body of apparently unrelated facts. This depends on an order in the universe that can allow understanding. One may choose, with Spinoza, to use the word God to name that Order. But this does not come anywhere close to the God of religious faith in the sense of Judaism or Christianity or Islam, who cares for us and has something to do with our life.

In 1980, I was invited to give the Gifford Lectures in Natural Theology at the University of Edinburgh. I had come to the brain as both a computer scientist and someone interested in psychological functions, so the challenge for me was to try and relate what I knew about the mind and the brain to theology. I did an immense amount of reading in theology and the history of the Christian church.

I was given a co-lecturer, Mary Hesse, who was at the time the Professor of History and Philosophy of Science at Cambridge, and who is a believing member of the high church of England. Although Mary and I agreed to differ on such "small matters" as free will and the existence of God I think I gained a deep understanding of the very real questions that motivated religious belief in general and theologians in particular.

One way we formulated our view of religious belief was to ask a slightly whimsical question: "Is God more like embarrassment or gravitation?" Embarrassment is a physiologically real phenomenon, but we understand that different societies have totally different constructs as to what it is to be embarrassed. There's no great embarrassing thing with a capital E out there that is the trigger for embarrassment.

Embarrassment is socially conditioned, and yet it does depend on certain basic factors in human physiology. On the other hand, although our theory of gravitation has changed from the Greeks to Newton to Einstein, there's still the feeling that there is an inherent phenomenon that has nothing to do with human society. If there were no humans there would be no embarrassment. But even if there were no humans there would still be gravitation. If you accept the first analogy you would see God as being a human social construct, but if you accept the second view, then God exists whether humans are here or not.

GS: An awful lot of modern neuroscience and modern philosophy and linguistics borrow metaphors and models from computer science. Does applying the computer as a model for the brain have religious ramifications?

MA: Different generations have different metaphors. In the 1920s and 1930s the brain was a telephone network. In the nineteenth century a lot of our understanding of the human was driven by the comparison of the human body with steam engines. It's always very easy to slip from saying "in certain ways, as I will carefully delineate, A is like B" to just saying "A *is* B."

In terms of ethics, people can find ways of coupling their view of the brain to their ethical view, but there's no real causal connection. For myself, although I'm not a believer in God, I certainly am a believer in a certain type of tolerance, a certain type of democratic set of institutions, respect for others, fair play. My growing knowledge of the brain just makes me increasingly aware of the beautiful, intricate subtlety of each human being, and makes me think that we should work for a society in which each of these exquisite machines can recognize its full potential.

GS: I'm wondering if your life as a Jew affects either what you study or your methodology in ways that you are aware of?

MA: I don't think so. Apart from the fact that I believe there are laws to be discovered, and the fact that I have some ethical standards, I don't see any deeper connections. If I go back to the Ten Commandments, then one might say that they're consistent with most of my ethical views. But I could just as well start from the Hippocratic Oath and ask how it migrates from being a doctor to being a researcher of biological process.

GS: Wittgenstein said, "We feel that when all scientific questions have been answered that the problems of life remain completely unanswered." I'm wondering if you find any truth in that statement and if you think that the science that you've seen other religious scientists pursue has left their religious longings and religious needs pretty much intact?

MA: Certainly, some scientific achievements have relevance to social issues: some have contributed to the amelioration of pain, new medical techniques have restored people to life, chemical substances like Prozac can take a severely depressed person and make them an apparently well person again, while understanding the ecosystem better lets us suggest legislation and technological innovations that will allow us to preserve aspects of nature that we value. I don't see these as de-coupled from life at all. I think there can be a strong coupling between science and our social concerns and welfare.

In other words, science can provide new insights which shape the way we approach human questions. This does not mean that science by itself can answer all our human questions, so I do have a sense of what Wittgenstein is saying, but I wouldn't accept Wittgenstein's quote in its absolute form. I don't think the monks in the desert, the anchorites, by their total devotion to religion, were able to answer all of life's ques-tions either. We're finite beings and no matter what we do, there are going to be things that weigh upon us that we cannot address from our own perspective alone.

Michael A. Arbib

The horrors of humanity and the computation of the self

EGOTIST, n. A person of low taste, more interested in himself than in me.

FREEDOM, n. Exemption from the stress of authority in a beggarly half dozen of restraint's infinite multitude of methods. A political condition that every nation supposes itself to enjoy in virtual monopoly.

HABIT, n. A shackle for the free.

HEAVEN, n. A place where the wicked cease from troubling you with talk of their personal affairs, and the good listen with attention while you expound your own.

(Ambrose Bierce, The Devil's Dictionary)

Introduction

My concern in this essay is with the notion of free will in the presence of evil and its relation to our evolving understanding of the person in computational terms. As Ambrose Bierce makes clear in his bracingly dyspeptic definitions in *The Devil's Dictionary*, freedom is in great part illusory. We are seldom free of the demands of others, nor they of ours, and we each accept, often unconsciously, a social contract which greatly constrains our courses of action.

The issue becomes whether there is a kernel of individual freedom that remains when the social constraints and the pressure of others are stripped away. If so, is this freedom compatible with a view of the brain as a wondrously complex computer, or does it depend on some human, possibly divine, spark that transcends our physical being? If not, why is it that many people still have the illusion of freedom? In either case, what are the values that inform the quest of so many for freedom, how relevant is religion to that quest, and to what extent will modern technology, and our growing linkage into world wide webs of computers and humans, enhance or diminish these values?

Theodicy and purpose

Within theology itself, a crucial problem is that of theodicy. Why is there evil in the world? A typical answer is that God gave us free will – and this makes evil possible because we would not have free will if we were not ever allowed to be nasty to each other. Since, as Bierce notes, life actually imposes a multitude of constraints, why could a just God not impose the constraint of being "nice"? One Judeo-Christian answer (cf. the route of God's development as traced in Miles 1995) might be that God was subject to temper tantrums, and so the *imago Dei* condemns us to uncontrollable passions, too. In the Old Testament God often punishes the chosen people because they have upset him; and even when rewarding them, God is doing terrible things to the Canaanites or some other non-chosen tribe. So one might say that the Old Testament anticipates theodicy by suggesting that God, in making a covenant with the Jews (some Christians say with the whole human race, others say only with those who have accepted Christianity), decrees that you do not enjoy divine blessing without accepting responsibility to God's wishes.

What, then, are we to make of the notion of a divine purpose? If you have a certain type of religious belief, you can rationalize the world's horrors away by saying this is a vale of suffering and that if somebody dies a horrible death, then surely it will make the sweetness of heaven all the greater. For people who can maintain such a belief system, it is an incredible source of strength to make up for what is wrong with the world. However, this is not evidence for a divine purpose, only an account of one way in which belief in such a purpose may be ameliorative.

But what I have just said has nothing to do with being a scientist, let alone a computer scientist. It is purely the perspective of someone who reads the newspaper or watches television news and ponders the broader implications of what he learns. How do I get back to the question of overarching purpose as a scientist? Most people who have studied the biology of evolution have learned that we should not think of it as a process directed toward humankind as the peak, but rather as a process whereby it became possible for humans to emerge from a history that goes right back to before there were even single-celled creatures. We do not have a causal theory of how that happened, but we do have a coherent theory of how it could have happened, constrained by what we learn from the fossil data, the geological record, studies of animal populations and animal husbandry, and new insights gleaned from molecular biology. But natural selection, as understood in this neo-Darwinian synthesis, is not a *telos*-bound process.

To what extent, then, can we – humankind – give purpose to this possible emergence of a global society (which even now pushes beyond the confines of the home planet)? If there is a God-given essence of the human at work in the world, a divine purpose, it requires a very selective approach to the evidence to discern it, seeing all that is good as the gift of God, and all that is bad as the result of humanity's turning its back on God. For me, rather, this reveals the ongoing interplay of biology and culture, of evolution without purpose. To what extent, if any, can we as individuals act responsibly to prevent the horrors the world has seen and help people lead more meaningful lives? Do we have the will to take responsibility for our futures? As computer scientists we can address this issue in (at least) two ways: To what extent can computer science, by contributing to attempts to view the mind-brain as a computer, contribute to our understanding of human will and social interactions? To what extent can computer science, by creating new technologies for computation and communication, change the social matrix in which humans find themselves? The present essay emphasizes the first of these two questions.

Machine metaphors for brains, minds, and persons

The machine metaphor for the human goes back centuries. Vaucanson (1709–82) made incredible automata, including a mechanical flute player, and a duck which ate and drank with realistic motions of head and throat, produced the sound of quacking, and could pick up corn meal and swallow, digest, and excrete it. People were intrigued by these automata and speculated as to whether humans might indeed be automata. Long before Vaucanson, René Descartes (1596–1650) was very much aware of automata and set the course of much of Western philosophy – with a look over his shoulder at the church – by saying that the body as automaton accounted for much of what we were, but that the soul was separated from the body, communicating with it via the pineal gland. In this century, in addition to building automata which can do complex physical things, we can also program computers to do much that before might have been seen as more mental in character, such as processing language, recognizing pictures, and playing chess at grand master level.

People acting in a mechanical non-spontaneous way are often called automata, but this begs the very question for which cognitive science seeks a positive answer: "Is the working of the human mind reducible to information processing embodied in the workings of the human brain?" i.e., is human spontaneity and intelligence a purely material phenomenon? One's opinion on this today can be a personal reaction, just like the optimist and the pessimist looking at the glass and saying that it is half full or half empty. Somebody who appreciates the current state of modeling the brain, or the body, can either look at how much can be explained in these terms, and therefore by faith – and it is by faith, not something we can rationally prove – believe that all mental phenomena can be explained in terms of brain and body. Someone else can look at many aspects of personality and personal experience that are not in any way captured in our current models and say, "This is why I still believe in dualism, that the soul is separate."

Many people say "the brain is a computer" and mean this in the sense of today's electronic machines. However, in my own work I have made much of the point that the brain is a computer which is very different from any computer we have yet succeeded in building (see, e.g., Arbib 1989: chapter 9). My view is that it is the job of computer science not only to build better technology today, but also to create a theory which encompasses the brain's very different adaptive, distributed, multi-level computations. A human brain comprises hundreds of regions, each region with (hundreds of) millions of cells, each cell with tens of thousands of connections, and each connection involving subtle neurochemical processes – which gives a great deal of room for processes shaped through evolution that will continue to mystify us for decades to come. To understand such complexity is to push computation far beyond anything we can imagine today.

My concern, therefore, is not to try to reduce the human to the status of our current machines, or of those we can envision for the near future, but rather to see how our notion of machines might change and grow over the coming decades as we better understand how the brain functions. My own approach, "schema theory,"[1] suggests ways in which the science of information and computation can be engaged in a dialogue with the humanities. It was such a dialogue that Mary Hesse and I shared in preparation for the Gifford Lectures in Natural Theology in Edinburgh in November 1983.

Despite the integration of the individual and the social in our thinking, significant differences emerged between us. For Mary, God was still to be seen as a transcendent reality, where I would see a God as a social construct. Turning to the theme of this essay, we differed about free will. I hold what we called the "decisionist" position, which claims that to have free will is simply to act on the basis of one's schemas, and sees the term "freedom" as useful at a social level of analysis: we are free if others do not unduly coerce or indoctrinate us. This contrasts with the "voluntarist" position, as we called Mary's viewpoint, which locates freedom in the self's decisions which transcend physical law.[2]

The voluntarist view of human freedom

The voluntarist says, "Surely, it is I who choose whom I marry or what I will eat at the restaurant. It cannot be that the current set of schemas and the state of the world can determine my choices, or the probability of those choices. How do you account for my sure knowledge that I have many times had to think long and hard to search my very soul before making agonizing choices that have drastically affected not only myself, but other people who are very important to me?" For the voluntarist, then, there must be the notion that human choice can break the chain of causality. Thus, in particular, voluntarism is incompatible with determinism. The decisionist does not either require or reject determinism. For the decisionist, it is simply enough that schema activation, which includes consciousness, involves processes that occur within spatiotemporal reality. The challenge for the decisionist is to see how these processes, instantiated within the brain, can still give us an experiential life which is diverse and rich in the way that we subjectively believe to be the case.

The crux for a voluntarist who wishes to engage in dialogue with the findings of science is to see how this freedom of choice could be compatible with – though not determined by – the laws of nature. It is suggested that the answer may be found in quantum mechanics, which is not deterministic; rather, the future is only predictable at the level of probabilities, and these uncertainties, these probabilities, can affect the macro-world. A random movement of an electron can be amplified to throw a switch one way or the other. Membrane noise in a neuron can determine whether or not a neuron reaches threshold and this can affect some mental state. The voluntarist will then accept the quantum-mechanical view of physics as being our current "best bet" about how the world works, and then build upon the claim that quantum mechanics only determines a set of possible future states of the world to claim that the unique individual can transcend spatiotemporal reality to choose within that ensemble of possibilities.

Notice that this view does not crudely reduce human choice to the random tossing of a quantum-mechanical coin. We are not free, says the voluntarist, because we are random, but rather because randomness leaves us room in which to make individual choices. In other words, this element of choice is compatible with, but goes beyond, physical causality. The voluntarist thus regards the mind-brain essentially as an open system, in terms of which some events are simply uncaused at any physical level, and so needs some supplementary account of free will in terms of entities that are not physical, not in space and time. In our progress toward Edinburgh, Mary Hesse argued that the essence of the human was an emergent property, compatible with physical causality, but not entailed by it or any sub-personal development of it.

Two views of emergence

Elsewhere (Arbib 1991) I have discussed the debate between those who see emergent properties as transcending the properties of the constituents, no matter how rich their interactions might be, and those who seek to explicate these apparently new properties by a new and more subtle analysis of the collective impact of such interactions. I expect the "science of the person" to provide a satisfactory account of the self as a property that is emergent in the latter sense from the individual's body, brain and schemas, with no appeal to a transcendent self of the kind made by Hesse.

More explicitly, an emergent property is one that we can observe in a complex structure that we cannot observe in the isolated pieces. With a statistical averaging assumption, we can understand many emergent properties of a fluid or a magnet in terms of a theory of the underlying "pieces," a physical theory of the statistical mechanics of molecular interactions. I hold, as the working hypothesis for my research in cognitive neuroscience, that mental properties are of the same kind, though this has certainly not been proved. I argue that, though it will require developments in the underlying theory, we will be able to explain all mental phenomena in naturalistic terms. I espouse, then, a cognitive neuroscience which approaches the mental by seeking to place the firing patterns of neurons within a coherent overall organization provided by a schema-theoretic account of an animal's behavior or a human's mental state.

When it comes to mind and brain, our understanding of the macro-level (language concepts, intentionality, conscious behavior) and the micro-level (the properties of neural circuits) are both at an early stage of development. We may expect them to change under the analysis of multiple levels. We must not only provide formal models of limited phenomena where we can (in computer terms, mathematical equations, or neural networks) but also understand the limitations of those accounts in describing the richness of human experience. In this way we can conduct a dialogue between the charted and the unknown. Scientific study of the brain will need both the "high-level" language of schema theory to describe mental states and behavior (and then we study cognitive psychology) and "low-level" language to describe the anatomy and physiology of neurons (call it brain theory).

By contrast, we may speak of the "strict" view of an emergent property as one that emerges as a new property for systems of sufficient complexity, which cannot in any sense be explained in terms of lower-level interactions. Hesse takes this view of the person as an emergent property. She cannot foresee a future development of the physical or the brain sciences which would come to explain what it was to be a person in terms of the interaction of schemas or neurons. It would require a new analysis of the specifically human which had to be seen as something that could emerge within a sufficiently complex mind-brain system, but which could never be deduced from the working of that system, no matter what the change of our laws of physics or physiology might be. In this way, there could be downward causation from this emergent structure, the mind-body complex. This "mind-body complex" – not a disembodied soul acting from outside upon the body – would constitute the person, and would affect the way neurons interact. Choice, decision, responsibility, praise and blame – all these predicates – would then only be explicable in terms of this emergent complex of the human person and could not, even in principle, be reduced to theories of the brain.

Mary Hesse and the neurophysiologist John Eccles both appeal to the Heisenberg uncertainty principle to allow a non-corporeal mind, separate from the brain, to nudge the quanta in the synapses to alter the brain's dynamics from the course it would otherwise have followed. I reject this view. The voluntarist is looking for some transcendent essence of the human, even if it is an emergent essence, while the decisionist sees human personality as arising from our complexity within space and time as conditioned by our human bodies and our interactions within society.

The decisionist view of human freedom

The decisionist instead sees the "you" that makes choices as constituted by the holistic net of schema interactions within your brain. But this highly personal network has a coherence (limited though we must admit it to be, recalling Freud), and it is this coherence of the network that we call the self. This "you" is explicable within space and time, within the network of schema interactions that is at the core of the decisionist view of the person and, thus, of the decisionist view of freedom. This self is, in my view, no more transcendent with respect to our brains and bodies than temperature is transcendent with respect to molecular motion.[3]

The decisionist can give a coherent view of human freedom and can integrate this into a secular worldview that does not require transcendent principles, whether those of God or those of Habermas's (1979) ideal speech situation. Needless to say, in the short space that remains, it will not be possible for me to completely replace thousands of years of the Judeo-Christian tradition! At most, I can offer a number of promissory notes. To preface the decisionist case, I would recall the individualistic or liberal conception of freedom (Partridge 1967: 221–225) that goes back, for example, to John Stuart Mill's *Treatise on Liberty*. On this view, one is free to the extent that one is free from coercion, able to set one's own goals, and to choose between alternatives. One is free to the extent that one is not compelled to act as one would not choose to act, whether that compulsion comes from the will of another human or from the state. In analyzing this freedom, the liberal view sees that there are different forms of coercion. There can be direct forms of coercion in terms of commands or prohibitions backed up by force or superior power; but there are also indirect forms of coercion that mold the range of alternatives that are even considered, whether through propaganda or education. These latter are forms of coercion that do not restrict the choice already made, but rather restrict the wherewithal to consider alternatives and then to make a deliberate or informed choice. Literacy and freedom of the press are thus parts of this liberal conception of freedom because they provide the ability to express diverse viewpoints so that people can make their choices within an understanding of a rich repertoire of alternatives.

The decisionist view, in so far as it is a view of freedom (as distinct from an approach to cognitive science), is really a version of the liberal view. Much of such freedom may be illusory if it rests on socially conditioned schemas for which you have engaged in no conscious critique – and here we face an infinite regress, since the grounds of critique must themselves be critiqued, and on and on until we accept certain views as "obvious" or "God-given." There is no guarantee that our "free" choices will not seem monstrous and evil to others who hold different worldviews.

To see how the decisionist explains why we may have to think long and hard before making agonizing choices, I appeal to the notion of computational complexity. The fact that a computer program is deterministic (but note that the schema view is not necessarily deterministic) does not deny that it may take many hours or even days of computing before the computer will deliver an adequate answer. In the same way, relying on the computations of our schemas within the brain, we can be aware of many alternatives that have to be considered before reaching a decision. We may be aware that we have no easy rules, no preset evaluations, which can quickly determine the best alternatives. We may not even be sure what constitutes "best." We may also have some sense of the costs of a wrong decision. A life, a career, or the love of someone dear to us may hang in the balance on the basis of this decision.

The belief that decisions can be made through the working out in space and time of our schemas which are but imperfect reflections of natural and social complexity does not say that there is any quick shortcut to computation. And so we, aware of these complexities, may find ourselves forced to make a decision knowing that we have imperfect data, knowing that we have too little time, knowing that the outcome of the decision is of immense importance to us. And here, it seems to me, is a sense of "thinking long and hard to make agonizing choices" that requires nothing transcendent for its understanding. Much of the decision making that we engage in, whether or not it is conscious, is a time-limited process of analogical reasoning, rather than the completely elaborated, certain deduction from consistent axioms.

Going further, the voluntarist has a rich vocabulary with which to talk about the phenomenology of our everyday experience, including such terms as "feeling guilt," "accepting responsibility," and "deserving praise or blame." Let me attempt a decisionist account of a couple of these in a rather improvised way as an invitation to further explore whether the language of schema theory can indeed bring us to some sort of closure in giving an account of subjective experience.

To analyze "feeling guilt," we can go back to Freud's discussion of the superego and then reflect it forward into schema-theoretic terms (cf. Arbib and Hesse 1986: chapter 6): schemas can embody our sense that there are socially sanctioned courses of action, but also an awareness that our own schemas are not simply devices for following the social consensus. We may then be aware that in a particular situation the latter may lead us to make choices which are discordant with those we believe society or our parents would ask of us. In some cases, we are able to call upon our stock of schemas to rationalize our course of action, but in other cases we must confess to a weakness or impulse that caused us to act in a way that we cannot choose to defend. It is this notion of indefensibility that might provide the core of a decisionist account of guilt. All those who have ever been on a diet will know what I am talking about!

In the same way we can talk about notions of responsibility. In fact, the sense of responsibility may be more pressing for the decisionist than for the believer in a soul: if you believe that there is a soul that provides a pipeline to God, then responsibility reduces to whether or not you open the communication channel. From a decisionist viewpoint, we may say that you are a responsible member of the society not so much if, in a standard situation, you emit the stock response, but rather if, finding yourself in a new and complex situation, you can nonetheless marshal your schemas in such a way that you act in some way that others, with time and leisure to reflect, would agree to be appropriate.

Punishment at any time is something that is historically rooted. We may provide a critique of it, grounded either in transcendental principles or in terms of social mechanisms. Having done that, we may either wish to preserve a tradition, or we may change the form of punishment in terms of its effectiveness either as a deterrent or as an example to others. A term like "punishment" thus comes to be seen as shorthand for a variety of complex situations that we can analyze at great length. But what then of a phrase like "he deserves punishment"? A decisionist might say that it is shorthand for some expression like: "Here is a person who in many situations has shown himself capable of evaluating certain alternatives. Here is a situation in which the person did not choose to avoid some grievous situation. We have no way of explaining how that choice was made, save to see it as a failure of certain accepted principles of choice on that occasion. For that reason, the person will be punished." Clearly, to the extent that we come to accept such shorthand, to that extent will the affective force of many terms in our current moral vocabulary be altered. However, it would be wrong to think that any single decisionist expansion of the above kind could exhaust the meaning that such terms have as part of our personal reality.

When we talk about diminished responsibility in a court case, we are simply trying to evaluate whether the mind of the defendant is such that he has the wherewithal, given a situation, to come to a "responsible" decision. Much of the law – as in distinguishing manslaughter from murder, or arguing for diminished responsibility – can be seen as depending on the attempt by lawyers to construct a decisionist theory of the person. The impact of the law is to be judged, not on the basis of a transcendent entity (though such axioms as "all men are created equal" are in some sense Enlightenment extracts from a transcendental Christian view of the soul), but in terms of an analysis of human decision making, worked out as we would say through our schemas, to give an account of the role of deterrence and punishment, relating social goals to those of the individual.

A secular schema for society

Computer technology and cognitive science have much to offer in forging human self-understanding for the coming age. This new technology allows us to make massive changes in workplaces, social structures, and our self-image. With what do we structure this endless array of choices? My task in this essay has been to suggest how one might build upon a schema theory limited to the more scientifically tractable areas of cognitive science in order to begin charting the density and complexity of human, moral experience with eyes open to the human potential for evil as much as for good. Yet I do not think that there is any model of our future society which is implied by schema theory. To see why this is so, consider two lessons of schema theory:

1   "Schema theory teaches us that people are inherently different because each person has his own stock of schemas." But if we try to infer from this a social view, are we to infer that this view must respect individuality, or are we to see the need for imposing tighter controls so that individuality will not damage the social consensus?

2   "Schema theory tells us that knowledge is provisional." But does this lead us to respect pluralism, or does it rather say that we should restrict inquiry to reduce the risk of destabilizing change?

Given these possibilities, I can offer a secular schema of society compatible with a decisionist, non-transcendental view of the person, but with no claim to certainty that it is "correct." It incorporates the notion that a person may be free if he chooses to follow custom, but not if he has been conditioned to always follow the changing dictates of a ruler. Freedom seems to require not so much that one must always choose to change, but rather the possibility of such choices, and for this the various alternatives must be known (Partridge 1967).

I thus join those who hold the liberal view of freedom in seeking a society in which a wide variety of beliefs are expressed and where there is a considerable diversity of tastes, customs, codes of conduct and styles of living. I would see this as more likely in a society in which power is widely distributed. My overarching secular schema, then, is one of a pluralistic, adaptive society which accepts that tradition may contain much wisdom of continuing relevance, yet sees that portions of the tradition are harmful or simply irrelevant in the modern world. Certain schemas – think of those that guide warfare – were acquired in a socio-historical context, such that they are inappropriate to be acted upon in a modern context. Even worse, the present context lacks the necessary feedback or social pressure to recalibrate them.

Many of the patterns of our daily lives, whether our own idiosyncrasies or the conventions of a language community or a social group, are purely human constructs. For example, in California we must drive on the right-hand side of the road; in New South Wales on the left. But this is a purely social convention: life is easier if the members of a single driving community observe the same basic rules. We do not hold that there is a greater Reality, the one true Road-Reality, which the framers of human road rules try to approximate. What then of human reality? Is there a God-reality that transcends space and time, and in which alone human life can find meaning? Are there rules of social development and historical process which inevitably shape human nature? Is our reality to be found in the free will of the individual, and is this essence of the human unconfined by the shackles of physical law? Or, despite all these alternatives, will our growing understanding of the human reshape again our concepts of spatiotemporal reality while letting us come to understand humans as biological mechanisms that must wrest their own meaning from the historical contingencies of biological and social evolution?

It is certainly challenging to explain how an atheist can believe in the "sanctity of human life," but such acceptance does not imply acceptance of God. There are many components of belief in God, and many believers accept different sets of these components. In declaring oneself an atheist, one rejects "most of the package," but one is nonetheless shaped in diverse ways by the social world in which one lives, and may certainly accept certain ethical positions associated with one or another religious tradition. Sociobiologists have developed models of altruism which, even if still debatable, certainly demonstrate the viability of a God-free explanation for your relative valuation of an animal, a human (perhaps differentially weighted for family, tribe, and other groupings), and yourself. Moreover, cultural evolution builds upon biological evolution – and it is, to me, much more plausible to see the variety of human moral systems in these terms rather than to postulate a God who directly created humans with a (lamentably imperfect) moral sense, or who was able to choose laws of nature in order that moral agents could evolve in this universe.

We find it useful to speak of gravitation as an "external reality" of which science offers descriptions which may change yet which are increasingly successful according to the pragmatic criterion. However, we have already seen that the choice of whether to drive on the right-hand or left-hand side of the road ("Road-Reality") is an arbitrary but necessary, and necessarily social, convention. Nonetheless, we can see the possibility of a social analysis which explains why such a convention becomes a necessity. This leads us not to a set of social axioms, but rather to a consideration of the multitude of social choices before us. We are faced with a network in which we seek a measure of social coherence, rather than a delimited set of phenomena for which we can subject closed models to a well-defined pragmatic criterion. I do not deny the value of a set of well-grounded principles to guide our interactions within a particular socio-historical context. But I assert that our critique of society must be an ongoing process rooted in lived experience.

What we need is not a set of axioms that all members of society should hold in totality to yield a consensus on all these values, but rather a society in which there are accepted patterns of communication which allow for moral discourse while not necessarily determining its outcome, a society that allows us to come to accept and live with the provisional nature of our personal and social knowledge. In the case of gravitation, we speak of universal laws; in the case of side-of-the-road choosing, we speak of a pure convention, yet hope to give an explanation of the need for such a convention.

Our ethics is defined by a set of conventions subject to individual and social critique, rather than being the subject of universal laws (no matter how dimly perceived) like gravitation.

From a different perspective, E.O. Wilson (1978), in developing the human implications of his sociobiology, has argued that "the good" is whatever evolution shows to be fittest in terms of natural selection. I would argue, as would the religious believer, that our moral intuitions may require us to resist the latest social consensus or even the ultimate biological consensus and may cause us to feel that we as humans have learned something that cannot be expressed within the blind processes of natural selection and evolutionary trends. Yet, unlike the theist, I resist any notion of ultimate principles, God-given or otherwise. An enlightened theist like Mary Hesse might respond as follows: "The way we read the Bible is part of a hermeneutic process. The Bible was written by men who believed that the end of the world was at hand. Thus, at the end of the first century, people had to learn to read the Bible in a new way, because the temporal world was enduring. When Augustine wrote at the Fall of Rome, he found a way of reading the Bible that gave comfort to men who saw the greatest works of civilization crumbling by offering instead the endurance of the city of God. Aquinas was shaped in great part by writing his *Summa Contra Gentiles*, his attempt to refute the teachings of Islam."

Thus there are religious believers who would say that, though the root of human existence is still to be found in our relation to God, the tradition is not something static. It is a body of wisdom that can and must adapt to the age. When one surveys the history of the Christian religion, one sees that Christian society has often been one of totalitarian and sectarian violence – the expulsion of the Jews from Spain, the Inquisition, anti-Semitism. There are also places and times in history in which religious belief has allied itself with the fullest expression of human freedom. In the present world, we can perhaps look at the support by many Roman Catholics (but not the hierarchy) for the liberation theology of Latin America. One can argue that it is an evolving sense of social justice, rather than religion, that is driving this movement. Witness the long-standing support of the church in Latin America, until recently, for the privileged rather than the oppressed. What makes one a secularist, rather than a religious believer, is that one cannot accept a transcendent view of salvation, but rather finds oneself engaged in an evolving critique of society as lived within this present world.

Conclusion

My view is that there are no ultimate axioms, no canonical books. Rather, we are continually reacting, changing, updating our views in response to historical processes. In this present world, we cannot look exclusively to the grounding of ancient principles, but rather must grapple with the new challenges of nuclear war, abortion, civil rights, the plight of refugees, damage to the environment, and the changing realities of family, friends and work-place. What becomes of the specifically human in such a secular view? All I can say is that we must adapt current data about personal experience as part of what is to be explained without denying that such experience may be open to change. Human values are not grounded in God or any other ulti-mate principle. Rather, we continually start anew from our current values and subject them to a critique which is itself history-laden.

The question, "What are the grounds of human values?" may be put in perspective by the apparently frivolous question, "What are the grounds of kangaroos?" There is no essence of the ideal animal that justifies the exis-tence of the kangaroo. Rather, we use neo-Darwinian theory to understand the evolution of the kangaroo within a rich historical and geographical context. It seems to me that this evolutionary view must also apply to human societies and to the religious texts that form an integral part of the culture of those societies. Religious texts offer many resonant metaphors that seem to help us in trying to chart the complexity of our experience, using such words as "evil," "salvation," "destiny." Are we to give up these words? Perhaps not, but I think we must defuse many of the metaphors that infuse them with their current meaning. We can no longer see evil as some absolute concept, whether it be embodied within the Devil or whether it be some principle with which God and the God-fearing are locked in eternal struggle. There are institutions and actions and people that we may agree are evil. Consider just one record from the contemporary history of the horrors of humanity. Mark Danner (1997) quotes many examples of sadistic behavior of Serbian guards engaged in "ethnic cleansing" in the Yugoslavia of the mid-1990s, and includes one excerpt, from *The Tenth Circle of Hell*, by Rezak Hukanovic, who speaks of his terrible experiences in the third person, referring to himself as "Djemo": After being attacked by guards, one prisoner

> stood up a little or rather tried to, letting out excruciating screams. He was covered with blood. One guard took a water hose from a nearby hydrant and directed a strong jet at [him]. ... [The prisoner's] cries were of someone driven to insanity by pain. And then Djemo saw clearly what had happened: the guards had cut off the man's sexual organ and half of his behind.

I do not believe we can come to understand such horrors by an appeal to some absolute essence of evil. Rather, we will see the many occurrences of evil as each to be fought on their own grounds, and we will see the better-ment of the human lot as something to be sought time after time, whatever the odds. But I will deny again that there are absolute criteria for a salvation to be sought in a God-reality outside human history.

Our problem is that we have not yet learned to be fully secular. Our society is pluralist, but how are we to be pluralist? The fact that we agree that a rich, free society is one that tolerates many viewpoints does not imply that we condone unbounded tolerance. I think that we are in a society that for all its pluralism has agreed that there are certain "don'ts" – ethnic cleansing, slavery and kicking dogs. American civil society may be in transition with respect to views on homosexuality, but abortion is still a matter of agonizing debate for which I see no grounds for immediate consensus one way or the other. There is no evidence that any one religious faith can settle these matters. Indeed, for many righteous belief is taken to justify the excesses of intolerance. Given this situation, what we need is not a set of axioms that all members of society should hold in totality to yield a consensus on all these values, but rather a society in which there are accepted patterns of communication which allow for moral discourse while not necessarily determining its outcome, a society which allows us to come to accept and live with the provisional nature of our personal and social knowledge.

However far the process of critique and consensus may go in a society, the construction and interaction of schemas will continue to create surprises. Evolutionary processes create new species which by their very existence provide new ecological niches and possibilities for the evolution of yet further species. Just as there is no single solution to the problem of adaptation in biological evolution, so there is no unique solution to the search for human values in a changing and complex world.

Notes

1 I use the term "schema" to denote the basic functional unit of action, thought and perception, a unit whose functionality is distributed – in the first instance – across the networks of the individual human brain. An assemblage of some of these schemas represents our current situation. Planning then yields a co-ordinated control program of perceptual and motor schemas which guides our actions. To make sense of any given situation we call upon hundreds of schemas in our current schema assemblage. Our lifetime of experience might be encoded in a personal "encyclopedia" of hundreds of thousands of schemas (cf. Arbib 1989: chapters 2 and 5).
2 The account of these views that follows is based in part on the fifth chapters of both Arbib (1985) and Arbib and Hesse (1986).
3 Of course this is a view whose proof within schema theory still requires much research to be done before we can reach the level of scientific confidence afforded by the explication of thermodynamics through statistical mechanics.

References

Arbib, Michael A. (1985) *In Search of the Person: Philosophical Explorations in Cognitive Science*, Amherst: University of Massachusetts Press.
—— (1989) *The Metaphorical Brain 2: Neural Networks and Beyond*, New York: Wiley-Interscience.
—— (1991) "Neurons, Schemas, Persons, and Society," in *Individuality and Cooperative Action*, edited by J.E. Earley, Washington, D.C.: Georgetown University Press, pp. 63–86.

Arbib, Michael A. and Mary B. Hesse (1986) *The Construction of Reality*, Cambridge: Cambridge University Press.

Bierce, Ambrose (1946) *The Devil's Dictionary*, in *The Collected Writings of Ambrose Bierce*, New York: The Citadel Press.

Danner, Mark (1997) "America and the Bosnia Genocide," *New York Review of Books* December 4: 55–65.

Habermas, Jürgen (1979) *Communication and the Evolution of Society*, translated by T. McCarthy, Boston: Beacon Press.

Hukanovic, Rezak (1996) *The Tenth Circle of Hell: A Memoir of Life in the Death Camps of Bosnia*, Foreword by Elie Wiesel, translated by Coleen London and Midhat Ridjanovic, edited by Ammiel Alcalay, New York: Basic Books.

Miles, Jack (1995) *God: A Biography*, New York: Vintage Books.

Partridge, Percy H. (1967) "Freedom," in *Encyclopedia of Philosophy*, edited by Paul Edwards, New York: Macmillan.

Wilson, E.O. (1978) *On Human Nature*, Cambridge: Harvard University Press.

13 Andrei Linde

Andrei Linde is a Professor of Physics at Stanford University. One of the authors of the inflationary universe scenario, which is the latest version of the Big Bang theory, he demonstrated that during the early stages of its evolution the universe may enter the stage of self-reproduction. At this stage the universe becomes divided into indefinitely large numbers of exponentially large regions with different physical laws in each of them. Linde has written 175 papers on particle physics, phase transitions and cosmology, as well as two books on particle physics and quantum cosmology. He has been awarded the Lomonosov prize of the Academy of Sciences of the USSR, and he was a Morris Loeb lecturer at Harvard University.

Interview by Philip Clayton

PC: I would like to start by asking about your exposure to the Western religious traditions at earlier points in your life, and the way that your training as a scientist might have affected or influenced that.

AL: I was educated in Russia, the Soviet Union at that time. The whole society was opposed to any type of religious education. It was not part of my university studies.

PC: Were religious influences present in an indirect way, in the background?

AL: Only the Christian religion. We were exposed to it a little bit by looking into books by Russian writers, some of whom were quite religious, and others who were fighting with religion. But it was not part of the culture in the Soviet Union. In fact, everybody safely survived with the assumption that it is silly to think about religion. Religion was a part of counter-culture in the Soviet Union but it was not a serious part of your internal life.

PC: Was this attitude more pronounced among scientists during your time of training?

AL: I didn't have much contact with non-scientists. In this sense, my experience was rather limited. As for scientists with whom I discussed philosophical things, they all had been very skeptical of the official dogma of dialectic materialism, which in our country was a kind of aggressive religion, because dialectical materialism was very, very careful in prescribing what scientists should think and how they should think, and what they should not do. This dogmatism raised an allergic reaction among scientists.

PC: Did it inoculate you against philosophical reflection?

AL: Not me, but this was the case for many scientists because they were exposed only to materialistic philosophy and they were exposed to this in a form of obligatory statements. It was necessary for all of us to pass exams in dialectical materialism. People, even with their ironic attitude toward materialism, did not have any reason to counter it. The generation of scientists which grew up during this time was completely materialistic.

The problem is that when you are talking about general relativity or quantum mechanics, you're asking questions which are on the border between science, philosophy, religion and something else. Take, for example, the special theory of relativity. Einstein discovered that two events may seem simultaneous to some observers and not simultaneous to other observers. This sounds absurd to many people. But after you realize that this is true, you'll think about it and wonder what kind of other absurd theories have a chance to be true.

PC: Einstein was willing to take some of his theoretical conclusions and speak of them in a more religious fashion, not in the sense of traditional religion, of course, but to say things like, God is sophisticated, but not deceitful. Is that an idea you also would be inclined toward?

AL: Sure. I sometimes use that kind of formulation, thinking and talking about inflationary theory which I co-invented. The inflationary theory is so simple, and simultaneously solved so many problems, that when I realized that this is the case, I thought that it would be absolutely stupid if God would not use this method of simplification of his own job.

PC: When you say it would be stupid for him not to do that, does the word 'God' function as a metaphor for saying that this is an extremely fruitful scientific theory, or is it really a religious statement?

AL: Well, mainly this is a joke, a metaphor. On the other hand, every joke has two sides, if the joke is good. After a really good joke, you'll sometimes think, is it really only a joke, or something more? The way science is organized is to try to cut off everything that cannot be explained, so that probably, in the end, you'll find out that everything is explained. You are not assuming that there is something that cannot be covered by simple laws of physics. I do not like this presumption. But nevertheless, the tendency of science is to go as far as we can without involving any hypotheses like that.

PC: So a basic assumption in doing physics, that we eventually can explain everything scientifically, means that we continually diminish the religious explanation?

AL: Yes, except that from the beginning this understanding of God as somebody sitting and creating the universe was from, well, bad movies. There are some primitive ideas in the minds of people until they have found something better. The question is, what is the deepest level of ideas about God that may survive after science investigates things?

PC: Would you, then, be inclined to have some sort of realm of truths that were not fully scientific, some realm remaining for the religious or spiritual?

AL: I think that this area does exist, but traditional science doesn't want to recognize it. The idea of physics is that eventually it will have all its explanations in itself, but this is not what is actually occurring with the development of science. There are always windows for expansion. The main goal of science is to make it a closed science, i.e. a self-sufficient science explaining everything. But there is no theorem saying that closed science does exist.

There is one area where it might be manifestly non-closed, and people just do not want to look into it because they have good reasons to think that this is irrelevant for most of their purposes. This place is the description of consciousness.

Some scientists think that consciousness is just a manifestation of matter, not something separate – that my thought is just the motion of electrons in my brain. From my perspective this is a very naive attitude which is simply a consequence of being too focused on the success of the materialistic model of the world.

PC: How, then, would we make the movement back from a scientific model to that original consciousness?

AL: When I am trying to describe the world I must first start with things which are certain and real for me. My feelings, my consciousness, my pain. This is real. But psychologists would look at me and instead of my feelings they will study my reactions. They will say that I am in pain when I am crying, and that I am happy when I am smiling. One can build a computer that will smile when you touch it. But does it mean that the computer is happy? Does it feel or does it only react? Any scientific investigation that does not consider seriously questions like that misses an important part of nature.

PC: The primary part?

AL: Yes. The question is, is it possible to develop a scientific approach that would not miss it. I hope so. But this is not something that you can easily find under discussion at any scientific conferences.

PC: Would this distinction apply equally to the physical sciences, the biological sciences, and the so-called human sciences?

AL: Yes, absolutely.

PC: Does your interest in the religious philosophies of India provide a speculative framework out of which you tend to work in looking at these questions?

AL: When I was studying Indian philosophy, I was extremely excited that what they say sometimes is painfully close to what I think. With many parts of Indian philosophical thought, you never know whether it is an allegorical way of expressing things, or whether it is literally what people thought at that time. If you are exposed to these archetypes of thought, then you can use them in your scientific work without taking them literally.

PC: It seems to me that you've sketched a positive, bi-directional interaction between two different fields: the field of scientific work, and the field of philosophical or religious reflection.

AL: I am trying to use physics and science in general to see clearly the boundary of physics and science, and I am attempting to investigate what is beyond this boundary. This is a way toward understanding something greater than simply the world of matter. But I would stop short from being tempted to formulate it as a way to study deity. This is something that I am opposed to.

PC: You're denying that science helps us to understand the nature of God?

AL: Yes, I do not like this formulation. I was brought up in Russia where we first believed that Lenin is a great person who made all people free. Then there was a generation of people who believed that Stalin is a great person who made all of us free. Now, monotheistic religions insist that there is but one God, and he came to make all of us free. I know that Stalin did not have the slightest intention of making us free. When I am thinking about the idea that there is one person, however great and nice and sweet he would be, who would come in order to make all of us free, I always remember how dangerous this kind of belief can be. When I see how rapidly some people in Russia are switching from communism to religion, I always wonder whether they are looking for freedom or for a different kind of slavery.

PC: Do you have a different response to poetic language? Is there a realm of the non-scientific, of the poetic or aesthetic, in which such religious thoughts and feelings might be safely expressed?

AL: If you read the Bhagavad Gita, for example, you can enjoy poetic language that existed for thousands of years. But there were so many different poetic traditions. One should be careful not to overemphasize one of them and harm some others. Also, one should carefully consider whether this is simply poetic or metaphoric language or something else.

PC: How about the realm of, for lack of a better word, the mystical? Would you allow for a rigorous scientist to have a mystical response to the universe that he studies, a sense of something deeper, beyond words?

AL: If truth becomes obvious to a person, then there is a chance that this is really a truth. The question is whether he can explain and transfer it to other people so that they would see the same. You may try to use a religious inspiration in your work as a tool. The question is whether it is only a tool, to be forgotten after you obtain reliable scientific results, or something more?

Andrei Linde

The universe, life and consciousness

Introduction

For many years, cosmologists believed that the universe began like an expanding ball of fire. This explosive beginning of the universe was called the Big Bang. Twenty years ago a different scenario of the universe's evolution was proposed. The main idea of the new scenario was that the universe at the very early stages of its evolution went through a stage of "inflation," a period of exponentially rapid expansion in an unstable vacuum-like state. As a result, the universe became exponentially large within a very short time. At the end of this inflationary period the energy of the vacuum-like state rapidly transformed into thermal energy, the universe became hot, and its subsequent evolution followed the standard Big Bang theory.

In many versions of the inflationary universe scenario, exponentially expanding parts of the universe permanently produce other exponentially expanding parts. Expansion ends in some of these parts and continues in others. Instead of the universe looking like a single expanding ball created in the Big Bang, we envisage it now as a huge growing fractal consisting of many inflating balls producing new balls, producing new balls, *ad infinitum*. In this essay, I will describe this new scenario of the evolution of the universe, along with some related conceptual problems.

The standard Big Bang model

According to the standard Big Bang theory, the universe was born at some moment $t = 0$ about fifteen billion years ago, in a state of infinitely high temperature T and density (the so-called "cosmological singularity"). Of course, we cannot really speak about infinite temperature and density. It is usually assumed that the standard description of the universe in terms of space and time becomes possible when its density drops below the so-called Planck density. The temperature of the expanding universe gradually decreased, and finally it evolved into the relatively cold universe we now live in.

This theory was extremely successful in explaining various features of our world. However, fifteen years ago physicists realized that it is plagued with a number of problems. For example, standard Big Bang theory being combined with the modern theory of elementary particles predicts the existence of a large amount of super-heavy stable particles carrying magnetic charge: magnetic monopoles. These objects have a typical mass 10^{16} times that of the proton. According to the standard Big Bang theory, monopoles should appear at the very early stages of the evolution of the universe, and they should now be as abundant as protons. In that case the mean density of matter in the universe would be about fifteen orders of magnitude higher than its present value. This forced physicists to look more attentively at the basic assumptions of the standard cosmological theory.

Problems with the standard model

The main problem of the Big Bang cosmology is the very existence of the Big Bang. One may wonder what was before the Big Bang? Where did the universe come from? If spacetime did not exist for times less than $t = 0$, how could everything appear from nothing? What appeared first: the universe or the laws determining its evolution? When we were born, the laws determining our development were written in the genetic code of our parents. But where were the laws of physics written when there was no universe?

This problem of cosmological singularity still remains the most difficult problem of modern cosmology. However, as we will see soon, we can now look at it from a totally different perspective. Let us now ask a much simpler question. At school we are taught that two parallel lines never cross. However, general relativity tells us that our universe is curved. The universe may be open, in which case parallel lines diverge from one another, or it may be closed, and then parallel lines cross each other like meridian lines on a globe. The only natural length parameter in general relativity is the Planck length $l_p \sim 10^{-33}$cm. Therefore one would expect our space to be very curved, with a typical radius of curvature about 10^{-33} cm. We see, however, that our universe is just about flat on a scale of 10^{28} cm, the radius of the observable part of the universe. The results of our observations differ from our theoretical expectations by more than sixty orders of magnitude!

Let us ask another naive question: why are there so many different people on the Earth? Well, the Earth is large and can accommodate many people. But why is the Earth so large? In fact, it is extremely small as compared with the whole universe. Then why is the universe so large? Let us consider a universe with a Planck density just emerging from the Big Bang. One can calculate how many particles such a universe would contain. And the answer is rather unexpected: the whole universe should contain just one particle, maybe ten particles, but not 10^{88} particles which are contained in the part of the universe which we see now. This is a contradiction by eighty-eight orders of magnitude.

The standard assumption of the Big Bang theory is that all parts of the universe began their expansion simultaneously, at the moment $t = 0$. But how could different parts of the universe synchronize the beginning of their expansion if they did not have any time for it? Who gave the command?

On a very large scale our universe is extremely homogeneous. On the scale of 10^{10} light years the distribution of matter departs from perfect homogeneity by less than one part in a hundred thousand. For a long time nobody had any idea why the universe was so homogeneous. Those who do not have good ideas sometimes have principles. One of the cornerstones of standard cosmology was the "cosmological principle," which asserts that the universe must be homogeneous. However, this does not help much since the universe contains stars, galaxies and other important deviations from homogeneity. We have two opposite problems to solve. First of all, we must explain why our universe is so homogeneous, and then we need to suggest some mechanism for producing galaxies.

All these problems are extremely difficult, which is why it is encouraging that most of these problems can be resolved in the context of one simple scenario of the evolution of the universe – the inflationary scenario.

Scalar fields and the uniqueness problem

In order to explain basic features of inflationary cosmology, we first need to make an excursion into the theory of elementary particles. Rapid progress of this theory during the last two decades has become possible with the unification of weak, strong and electromagnetic interactions.

It is well known that electrically charged particles interact with each other by creating an electromagnetic field around them. Small excitations of this field are called photons. Photons do not have any mass, which is the main reason why electrically charged particles can easily interact with each other at a very large distance. Scientists believe that weak and strong interactions are mediated by similar particles. For example, weak interactions are mediated by particles called W and Z. However, whereas photons are massless particles, the particles W and Z are extremely heavy; it is very difficult to produce them. That is why weak interactions are so weak. In order to obtain a unified description of weak and electromagnetic interactions despite the obvious difference in properties of photons and the W and Z particles, physicists introduced scalar fields, which will play the central role in our discussion.

The theory of scalar fields is very simple. The closest analog of a scalar field is the electrostatic potential Φ. Electric and magnetic fields E and H appear only if this potential is inhomogeneous or if it changes in time. If the whole universe would have the same electrostatic potential, say, 110 volts, then nobody would notice it; it would be just another vacuum state. Similarly, a constant scalar field φ looks like a vacuum state; we do not see it even if we are surrounded by it.

The main difference is that the constant electrostatic field Φ does not have its own energy, whereas the scalar field φ may have potential energy density $V(\varphi)$. If $V(\varphi)$ has one minimum at $\varphi = \varphi_0$, then the whole universe eventually becomes filled by the field φ_0. This field is invisible, but if it interacts with particles W and Z, they become heavy. Meanwhile, since photons do not interact with the scalar field, they remain light. Therefore we may begin with a theory in which all particles initially are light, and there is no fundamental difference between weak and electromagnetic interactions. This difference appears later, when the universe becomes filled by the scalar field φ. At this moment the symmetry between different types of fundamental interactions becomes broken. This is the basic idea of all unified theories of weak, strong and electromagnetic interactions.

The existence of scalar fields contributes to what I call the "uniqueness problem": if the potential energy density $V(\varphi)$ has more than one minimum, then the field φ may occupy any of them.[1] This means that the same theory may have different "vacuum states," corresponding to different types of symmetry-breaking between fundamental interactions, and, as a result, to different laws of physics of elementary particles. To be more accurate, one should speak about different laws of low-energy physics. At an extremely high energy the difference in masses is no longer important, and the initial symmetry of all fundamental interactions again reveals itself.

Finally, many theories of elementary particles which are popular now assume that spacetime originally had considerably more than four dimensions, but that extra dimensions have been compactified, shrunk to a very small size. This explains why we cannot move in the corresponding directions, and why our spacetime looks four-dimensional. However, one may wonder why compactification stopped with four effective spacetime dimensions rather than two or five? Moreover, in the higher-dimensional theories compactification may occur in a thousand different ways. The values of coupling constants and particle masses after compactification strongly depend on the way compactification occurs. The difficulty of constructing theories which admit only one type of compactification and only one way of symmetry-breaking represents another aspect of the uniqueness problem.

The inflationary model

According to the Big Bang theory, the rate of expansion of the universe given by the Hubble "constant" H is large when the density of the universe is large. If the universe is filled by ordinary matter, then its density rapidly decreases as the universe expands. Therefore expansion of the universe rapidly slows down as its density decreases. This rapid decrease of the rate of the universe expansion is the main reason for all the problems with the standard Big Bang theory. However, because of the equivalence of mass and energy established by Einstein ($E = mc^2$), the potential energy density $V(\varphi)$ of the scalar field φ also contributes to the rate of expansion of the universe. In certain cases the energy density $V(\varphi)$ decreases much more slowly than the density of ordinary matter. This may lead to a stage of extremely rapid expansion (inflation) of the universe.

There are several different versions of inflationary theory. Let us consider the simplest model, which I called chaotic inflation. This model describes a scalar field φ with a mass m and with the potential energy density V(φ) = $m^2\varphi^2/2$. Since energy density has a minimum at φ = 0, one may expect that the scalar field φ should oscillate near this minimum. This is indeed the case if the universe does not expand. However, one can show that in a rapidly expanding universe the scalar field moves down very slowly, as a ball in a viscous liquid, viscosity being proportional to the speed of expansion.

From here we have only one step to make in order to understand where inflation comes from. If the scalar field φ initially was large, its energy density V(φ) was also large, and the universe expanded very rapidly. Because of this rapid expansion the scalar field was moving to the minimum of V(φ) very slowly, as a ball in a viscous liquid. Therefore at this stage the energy density V(φ), unlike the density of ordinary matter, remained almost constant, and expansion of the universe continued with a much greater speed than in the old cosmological theory: the size of the universe in this regime grows approximately as e^{Ht}, where H is the Hubble constant.

To understand the situation at a more formal level one should analyze two equations which describe inflation in our model: $d^2\varphi/d^2t + 3H\,d\varphi/dt =$ $-V'(\varphi)$, and $H^2 = 8GV(\varphi)/3$. The second equation is a slightly simplified Einstein equation for the scale factor (radius) of the universe a(t); H is the Hubble constant, H = (da/dt)/a, G is the gravitational constant. The term 3H dφ/dt in the first equation is similar to the friction (viscosity) term in the equation of motion for a harmonic oscillator. One can show that if V(φ) is approximately constant during a sufficiently long period of time, the last equation has an inflationary solution a(t)~e^{Ht}.

This stage of self-sustained exponentially rapid expansion of the universe was not very long. In a realistic version of our model its duration could be as short as 10^{-35} seconds. When the energy density of the field φ becomes sufficiently small, viscosity becomes small, inflation ends, and the scalar field φ begins to oscillate near the minimum of V(φ). As any rapidly oscillating classical field, it loses its energy by creating pairs of elementary particles. These particles interact with each other and come to a state of thermal equilibrium with some temperature T. From this time on, the corresponding part of the universe can be described by the standard Big Bang theory.

The main difference between inflationary theory and the old cosmology becomes clear when one calculates the size of a typical inflationary domain at the end of inflation. Investigation of this question shows that even if the size of the part of inflationary universe at the beginning of inflation in our model was as small as $l_p = 10^{-33}$ cm, after 10^{-35} seconds of inflation this domain acquires a huge size ~ 1,000,000,000,000 cm! These numbers are model-dependent, but in all realistic models this size appears to be many orders of magnitude greater than the size of the part of the universe which we can see now, l ~10^{28}cm. This immediately solves most of the problems of the old cosmological theory.

Our universe is so homogeneous because all inhomogeneities were stretched $10^{1,000,000,000,000}$ times. Moreover, because the universe becomes enormously large the density of primordial monopoles becomes exponentially diluted. Even if it was a closed universe of a size 10^{-33} cm, the distance between its "South" and "North" poles after inflation becomes many orders of magnitude greater than 10^{28} cm. We see only a tiny part of the huge cosmic balloon. This is why the universe looks so flat and why nobody has ever seen how parallel lines cross. This is also why we do not need expansion of the universe to begin simultaneously in 10^{88} different causally disconnected, Plank-sized domains. One such domain is enough to produce everything which we can see now!

Inflation and the production of galaxies

Solving many difficult cosmological problems simultaneously by a rapid stretching of the universe may seem too good to be true. Indeed, if all inhomogeneities were stretched away, what about galaxies? The answer is that while removing previously existing inhomogeneities, inflation also created new ones. The basic mechanism can be understood as follows.

According to quantum field theory, empty space is not entirely empty. It is filled with quantum fluctuations of all types of physical fields. These fluctuations can be regarded as waves of physical fields with all possible wavelengths. If the values of these fields, averaged over some macroscopically large time, vanish, then the space filled with these fields seems to us empty and can be called the vacuum.

In the inflationary universe the vacuum structure is more complicated. Waves which have a very short wavelength "do not know" that the universe is curved; they move in all directions with a speed approaching the speed of light. However, inflation very rapidly stretches these waves. Once their wavelengths become sufficiently large, these waves begin "feeling" that the universe is curved. At this moment they stop moving because of the effective viscosity of the expanding universe with respect to the scalar field.

The first quantum fluctuations to freeze are those with large wavelengths. The amplitude of the frozen waves does not change, whereas their wavelengths continue to grow exponentially. In the course of the expansion of the universe more and more fluctuations become stretched and freeze on top of each other. At this stage one can no longer call these waves "quantum fluctuations." Most of them, the ones with exponentially large wavelengths, do not move and do not disappear from being averaged over long periods of time. What results is an inhomogeneous distribution of the classical scalar field φ which does not oscillate. It is these inhomogeneities which are responsible for density perturbations in our universe and for the subsequent appearance of galaxies.

Self-reproducing universe

Here we come to the central part of our story, to the theory of an eternally existing, self-reproducing inflationary universe. As I have already mentioned, one can visualize quantum fluctuations of the scalar field in an inflationary universe as waves, which first move in all possible directions, and then freeze on top of each other. Each freezing wave slightly increases the value of the scalar field in some parts of the universe, and slightly decreases this field in other parts of the universe.

Now let us consider those rare parts of the universe where these freezing waves always increased the value of the scalar field φ, persistently pushing the scalar field uphill, to greater and greater values of its potential energy $V(\varphi)$. This is a very strange and obviously improbable regime. Indeed, the probability that the field φ will make one jump up (instead of jump down), is equal to 1/2, the probability that the next time it also jumps up is also 1/2, so that the probability that the field φ will, without any special reason, make n consecutive jumps in the same direction is extremely small; it will be proportional to $1/2^n$.

Normally one neglects such fluctuations. However, in our case they can be extremely important. Indeed, in those rare domains of the universe where the field jumps high enough, exponential expansion begins with ever increasing speed. Remember that the inflationary universe expands as e^{Ht}, where the Hubble constant is proportional to the square root of the energy density of the field φ. In our simple model with $V(\varphi) \sim \varphi^2$ the Hubble constant H will be simply proportional to φ. Thus, the higher the field φ jumps, the faster the universe expands. Very soon those rare domains, where the field φ persistently climbs up the wall, will acquire a much greater volume than those domains which keep sliding to the minimum of $V(\varphi)$ in accordance with the laws of classical physics.

From this theory it follows that if the universe contains at least one inflationary domain of a sufficiently large size, it begins unceasingly producing more and more new inflationary domains. Inflation at each particular point may end very quickly, but there will be many other places that will continue expanding exponentially. The total volume of all inflationary domains will grow without end.

Eternal inflation implies that the universe as a whole is immortal. Each particular part of the universe may appear from a singularity somewhere in the past, and it may end up in a singularity somewhere in the future. However, there is no end for the evolution of the whole universe. The situation with the beginning is less certain. It is most probable that each part of the inflationary universe has originated from some singularity in the past. However, at present we do not have any proof that all parts of the universe were created simultaneously in a general cosmological singularity, before which there was no space and time at all. Moreover, the total number of inflationary bubbles on our "cosmic tree" exponentially grows in time. Therefore most of the bubbles (including our own part of the universe) grow indefinitely far away from the root of this tree. This moves the possible beginning of the whole universe into the indefinite past.

The laws of physics in different domains

Until now we considered the simplest inflationary model with only one scalar field φ. In realistic models of elementary particles there are many different scalar fields. For example, in the unified theories of weak and electromagnetic interactions there exist at least two other scalar fields, Φ and H. In some versions of these theories the potential energy density of these fields has roughly a dozen different minima of the same depth. During inflation these fields, like the field φ, jump in all possible directions due to quantum fluctuations. After inflation they fall down to different minima of their energy density in different exponentially large parts of the universe. Remember now that properties of elementary particles and the laws of their interaction depend on scalar fields. This means that after inflation the universe becomes divided into exponentially large domains with different laws of low-energy physics. It does not mean that the fundamental law governing our universe is not unique. The situation here can be understood if one thinks about three different states of water: it can be a solid, liquid or a gas. It is the same water, but these three states look quite different from each other and have strikingly different properties; fish can live only in liquid water.

If fluctuations are not too strong, scalar fields cannot jump from one minimum of their energy density to another. In this case the new parts of the inflationary universe remember the "genetic code" of their parents. However, if fluctuations are sufficiently large, mutations occur, and the "laws of physics" in new bubbles change from one bubble to another. In some inflationary models quantum fluctuations become so strong that even the effective number of dimensions of space and time can change. According to these models, we find ourselves inside a four-dimensional domain with our kind of physical laws not because domains with different dimensionality and with different particle properties are impossible or improbable, but simply because our kind of life cannot exist in other domains.[2]

Inflation and creation

The new theory has irreversibly changed our understanding of the structure and fate of our universe. Simultaneously, it is changing our ideas about our own place in the world. It is interesting to ask whether the new picture of the world is compatible with common religious beliefs.

With the expansion of science it has become more and more complicated to talk about God in simplistic terms. Apparently, the laws of the universe work so precisely that we do not need any hypothesis of divine intervention in order to describe the behavior of the universe as we know it. However, one point in particular has remained hidden from science: the moment of creation of the universe as a whole. The creation of everything from nothing could seem to be too great a mystery for science to consider.

With the development of inflationary cosmology the situation has changed somewhat. The possibility that the universe eternally re-creates itself in all its possible forms does not necessarily solve the problem of creation, but pushes it back into the indefinite past. By doing so, the properties of our world become totally disentangled from the properties of the universe at the time when it was born (if there was such time at all). In other words, one may argue that the properties of our world do not represent the original design and cannot carry any message from the Creator.

Is this true? Is this the final word of science? I believe it is too early to say anything definite. In such a situation one may try to say something allegorical, which may turn out to be a joke, or may contain something more. Thus I would like to finish this discussion by considering three puzzling possibilities.

Is there a message in the laws of nature?

Recently there has been some discussion of whether it is possible to create a universe in a laboratory. Indeed, one would need only a milligram of matter in a vacuum-like exponentially expanding state, and then the process of self-reproduction would create from this matter not one universe but infinitely many!

It is still not quite clear whether this process is theoretically possible or technologically feasible, but let us imagine for a second that the answer to both of these questions is positive. Should we really try to build a new universe in a laboratory? How would one be able to use it? Indeed, one cannot "pump" energy from the new universe to ours, since this would contradict the conservation of energy. One cannot jump into the new universe, since at the moment of its creation it is microscopically small and extremely dense, and later it decouples from our universe. One even cannot send any information about oneself to those who will inhabit the new universe. If one tries, so to speak, to write down something "on the surface of the universe," then, in the coming billions of billions years, the inhabitants of the new universe will live in a corner of one letter. This is a consequence of a general rule: all local properties of the universe after inflation do not depend on initial conditions at the moment of its formation. It quickly becomes absolutely flat, homogeneous and isotropic, and any original message "imprinted" on the universe becomes unreadable.

We have been able to find only one exception to this rule. As I already mentioned, if chaotic inflation starts at a sufficiently large energy density, it goes on forever creating more and more inflationary domains. These domains contain matter in all possible "phase states" (or vacuum states), corresponding to all possible minima of the effective potential and all possible types of laws of physics compatible with inflation. However, if inflation starts at a sufficiently low energy density, as is often the case with the universes produced in a laboratory, then no such diversification occurs; inflation at a relatively small energy density does not change the symmetry-breaking pattern of the theory and the way in which spacetime is compactified. Therefore it seems that the only way to send a message to those who will live in the universe we are planning to create is to encrypt information into the properties of the vacuum state of the new universe, i.e. into the laws of the low-energy physics. One might achieve this by choosing a proper combination of temperature, pressure and external fields, which would lead to the creation of a universe in a desirable phase state.

The corresponding message can be long and informative enough only if there are extremely many ways of symmetry breaking and/or patterns of compactification in the underlying theory. This is exactly the case, e.g., in the superstring theory (this was for a long time considered to be one of the main problems of this theory). Another requirement for the informative message is that it should not be too simple. If, for example, masses of all particles were equal to each other, all coupling constants would be one and the corresponding message would be too short. Perhaps, one could say quite a lot by creating a universe in a strange vacuum state with the proton being 2,000 times heavier than the electron, W bosons being 100 times heavier than the proton, etc., i.e. in the vacuum state in which we live now. The stronger the symmetry-breaking, the more "unnatural" the subsequent relations between parameters of the theory, the more information the message can contain. Is this the reason why relations between particle masses and coupling constants in our universe look so bizarre? Does this mean that our universe was created by a physicist hacker? Does this mean that only physicists can read the message of God?

The universe and life

The possibility described above represents an ultimate example of the arrogance of science. However, is it conceivable that our understanding of the universe is too simplistic in a different way? Is it possible that we are making a conceptual mistake in assuming that the universe is real, and that it encompasses everything? It is very hard to answer this question, but we may at least try to examine it attentively. A good starting point is quantum cosmology, the theory which tries to unify cosmology and quantum mechanics.

If quantum mechanics is true, then one can try to find the wave function of the universe. This would allow us find out which events are probable and which are not. However, this often leads to problems of interpretation. For example, at the classical level one can speak of the age of the universe t. However, the essence of the Wheeler–DeWitt equation, which is the Schrödinger equation for the wave function of the universe, is that this wave function *does not depend on time*, since the total Hamiltonian of the universe, including the Hamiltonian of the gravitational field, vanishes identically. This result was obtained in 1967 by the "father" of quantum cosmology, Bryce DeWitt. Therefore one is in trouble if one wishes to describe the evolution of the universe with the help of its wave function: *The universe does not change in time*. It is immortal, and it is dead.

The resolution of this paradox is rather instructive. The notion of evolution is not applicable to the universe as a whole since there is no external observer with respect to the universe, and there is no external clock as well which would not belong to the universe. However, we do not actually ask why the universe *as a whole* is evolving in the way we see it. We are just trying to understand our own experimental data. Thus, a more precisely formulated question is, *why do we see* the universe evolving in time in a given way? In order to answer this question one should first divide the universe into two parts: (i) an observer with his clock and other measuring devices and (ii) the rest of the universe. Then it can be shown that the wave function of the rest of the universe does depend on the state of the clock of the observer, i.e. on his "time." This time dependence is in some sense "objective," which means that the results obtained by different (macroscopic) observers living in the same quantum state of the universe and using sufficiently good (macroscopic) measuring apparatus agree with each other.

Thus we see that by an investigation of the wave function of the universe *as a whole* one sometimes obtains information, e.g. that the universe does not evolve in time, which has no direct relevance to the observational data. In order to describe the universe *as we see it* one needs to divide the universe into several macroscopic pieces and then calculate a conditional probability in order to observe it in a given state under the obvious condition that the observer and the measuring apparatus do exist. Without introducing an observer, we have a dead universe, which does not evolve in time. Does this mean that an observer is simultaneously a creator?

Although this problem has been around for more than thirty years, it was easy to ignore. Indeed, we know that the universe is huge, and quantum mechanics is important only for the description of extremely small objects, such as elementary particles. Therefore, for all practical purposes one could forget about subtleties of quantum mechanics applied to the universe: there was no real need to apply quantum mechanics to the universe in the first place.

However, in the context of inflationary cosmology the situation is entirely different. Indeed, we believe now that galaxies emerged as a result of small quantum fluctuations produced during inflation. The universe itself could originate from less than one milligram of matter compressed to a size billions of times smaller than a size of an electron. Its different parts were formed during the quantum-mechanical process of self-reproduction of the universe. One may consider our part of the universe as an extremely long-living quantum fluctuation. In such a situation the interpretation of quantum mechanics becomes absolutely essential for the further progress in cosmology.

Let us remember the famous Schrödinger cat paradox. Suppose that we have a cat in a box, and its state (dead or alive) depends on quantum-mechanical chance. According to the Copenhagen interpretation of quantum mechanics, the cat is neither dead nor alive until one opens the cage, observes the cat, and by this observation reduces its wave function to the wave function of either a dead or alive cat. It does not make any sense to ask whether the cat was really dead or really alive before one opened the cage.

This sounds like a joke. Common sense tells us that the cat is real, and it can be either dead or alive, but it cannot be half-dead. We are happy that quantum mechanics helps us to make a CD player, but we do not want to spend much time thinking about the interpretation of quantum mechanics so long as we can use the rules and get the right answer. So let us ignore this paradox; who cares about the cat anyway?

But after the invention of inflationary theory we must think about the universe described by quantum mechanics. Suppose that somebody asks you how the universe behaved one millisecond after the Big Bang. According to quantum mechanics, this is the wrong question to ask. Reality is in the eye of the observer, and there were no observers in the early universe. Of course we do not really need to know an exact answer. We only need to know a subset of the possible histories of the universe consistent with our present observations in order to predict the future. This is quite satisfactory from a purely pragmatic point of view, as long as one recognizes the limitations of science and does not ask too many questions. If we do not care about the cat, we do not really care about the universe. But then we end up not really caring about the reality of matter.

This example demonstrates the unusually important role played by the concept of an observer in quantum cosmology. Most of the time, when discussing quantum cosmology, one can remain entirely within the bounds set by purely physical categories, regarding an observer simply as an automaton, and not dealing with questions of whether the observer has consciousness or feels anything during the process of observation. This limitation is harmless for many practical purposes. But we cannot rule out *a priori* the possibility that by avoiding the concept of consciousness in quantum cosmology we have artificially narrowed our outlook. A number of authors have underscored the complexity of the situation, replacing the word *observer* with the word *participant*, and introducing such terms as a "self-observing universe." In fact, the question may come down to whether standard physical theory is actually a closed system with regard to its description of the universe as a whole at the quantum level. Is it really possible fully to understand what the universe is without first understanding what life is?

Let us remember an example from the history of science which may prove to be rather instructive in this respect. Prior to the advent of the special theory of relativity, space, time and matter seemed to be three fundamentally different entities. Space was thought to be a kind of three-dimensional co-ordinate grid which, when supplemented by clocks, could be used to describe the motion of matter. Special relativity combined space and time into a unified whole. But spacetime nevertheless remained something of a fixed arena in which the properties of matter became manifest. As before, space itself possessed no intrinsic degrees of freedom, and it continued to play a secondary, subservient role as a tool for the description of the truly substantial material world.

The general theory of relativity brought with it a decisive change in this point of view. Spacetime and matter were found to be interdependent, and there was no longer any question of which one was more fundamental. Spacetime was also found to have its own degrees of freedom, associated with structural perturbations of the metric–gravitational waves. Thus, space can exist and change with time in the absence of electrons, protons, photons, etc.; in other words, space can change in the absence of anything that had *previously* (i.e., prior to general relativity) been subsumed by the term *matter*.

There has been a more recent trend toward a unified geometric theory of all fundamental interactions, including gravitation. Prior to the end of the 1970s, such a program seemed unrealizable; rigorous theorems had been proven on the impossibility of unifying spatial symmetries with the internal symmetries of elementary particle theory. Fortunately, these theorems were sidestepped with the discovery of super-symmetric theories. In these theories all particles can be interpreted in terms of the geometric properties of a multidimensional super-space. Space ceases to be simply a requisite mathematical adjunct for the description of the real world, and instead takes on greater and greater independent significance, gradually encompassing all the material particles under the guise of its own intrinsic degrees of freedom. In this picture, instead of using space for describing the only real thing, matter, we use the notion of matter in order to simplify our description of super-space. This change in our picture of the world is perhaps one of the most profound (and least known) consequences of modern physics.

The universe and consciousness

Finally, let us turn to consciousness. According to standard materialistic doctrine, consciousness, like spacetime before the invention of general relativity, plays a secondary, subservient role, being considered just a function of matter and a tool for the description of the truly existing material world. Let us remember, though, that our knowledge of the world begins with perceptions, not with matter. I know for sure that my pain exists, my "green" exists, and my "sweet" exists. I do not need any proof of their existence, because these events are a part of me; everything else is a theory. Later we find out that our perceptions obey some laws, which can be most conveniently formulated if we assume that there is some underlying reality beyond our perceptions. This model of the material world obeying laws of physics is so successful that we too readily forget our starting point and come to think that matter is the only reality, and that perceptions are only helpful for its description. This assumption is almost as natural (and maybe as false) as our previous assumption that space is only a mathematical tool for the description of matter. In fact, we are replacing the *reality* of our feelings with a successful *theory* of an independently existing material world. And the theory is so successful that we almost never think about its limitations until we are forced to address those deep issues which do not fit into our model of reality.

It is certainly possible that nothing similar to the modification and generalization of the concept of spacetime will occur with the concept of consciousness in the coming decades. But the thrust of research in quantum cosmology has taught us that the mere statement of a problem which might at first glance seem entirely metaphysical can sometimes, upon further reflection, take on real meaning and become highly significant for the further development of science. At this point, I would like to take a certain risk and formulate several questions for which we do not yet have any answers.

Is it not possible that consciousness, like spacetime, has its own intrinsic degrees of freedom, and that neglecting these will lead to a description of the universe that is fundamentally incomplete? What if our perceptions are as real (or maybe, in a certain sense, are even more real) than material objects? What if my red, my blue, my pain, are really existing objects, not merely reflections of the really existing material world? Is it possible to introduce a "space of elements of consciousness," and investigate the possibility that consciousness may exist by itself, even in the absence of matter, just like gravitational waves, excitations of space, may exist in the absence of protons and electrons? Will it not turn out, with further scientific developments, that the study of the universe and the study of consciousness will be inseparably linked, and that, ultimately, progress in the one will be impossible without progress in the other? After the development of a unified geometrical description of the weak, strong, electromagnetic, and gravitational interactions, will the next important step not be the development of a unified approach to our entire world, including the world of consciousness?

All of these questions might seem somewhat naive, but it becomes increasingly difficult to investigate quantum cosmology without making an attempt to answer them. A few years ago it would have seemed equally naive to ask why there are so many different things in the universe, why nobody has ever seen parallel lines intersect, why the universe is almost homogeneous and looks approximately the same at different locations, why spacetime is four-dimensional, and so on. Now, when inflationary cosmology provided a possible answer to these questions, one can only be surprised that prior to the 1980s, it was sometimes taken to be bad form even to discuss them.

It is best not to repeat old mistakes, but instead forthrightly to acknowledge that the problem of consciousness and the related problem of human life and death are not only unsolved, but at a fundamental level are virtually unexamined. It is tempting to seek connections and analogies of some kind, even if they are shallow and superficial ones at first, in studying another great problem – that of the birth, life, and death of the universe. It may yet become clear that these two problems are not so disparate after all.

Notes

1   A few years ago it would have seemed rather meaningless to ask why our space-time is four-dimensional, why the gravitational constant is so small, why the proton is two thousand times heavier than the electron, etc. Now these questions acquired a simple physical meaning and can no longer be ignored. As we will see, inflationary theory may help us to answer these questions as well.
2   This new picture of the universe just described is a result of the work of many cosmologists in different countries. A detailed description of this picture can be found, e.g., in Linde (1990).

Reference

Linde, Andrei (1990) *Particle Physics and Inflationary Cosmology*, Chur, Switzerland: Harwood Academic Publishers.

14 Brian Cantwell Smith

Brian Cantwell Smith is Professor of Cognitive Science, Computer Science and Philosophy, Adjunct Professor of the History and Philosophy of Science, and Assistant Director of the Cognitive Science Program at Indiana University in Bloomington, Indiana. Before moving to Indiana in 1996 he was Principal Scientist at the Xerox Palo Alto Research Center (PARC) and Adjunct Professor of Philosophy at Stanford University. He was a founder of the Center for the Study of Language and Information at Stanford University (CSLI), and a founder and first President of Computer Professionals for Social Responsibility (CPSR). Smith's research focuses on metaphysics, epistemology and the foundations and philosophy of computing.

Interview by Gordy Slack

GS: Can you start by saying something about your religious background and experience?

BCS: I grew up as a member of the United Church of Canada. My father was a theologian, technically also an ordained minister, though he worked as an academic. Although I've worked in and been under the influence of the sciences, there's a fair amount of continuity between what I believe and his worldview: his sense of significance, his sense of what it is to be religious, and the theological presuppositions that I was given as a child.

My father's theology was radical in a lot of ways. Lots of people think that to be religious is to believe certain things: that God exists, that someday we'll go to heaven, etc. My father argued that the tendency in the modern Western church to reduce being religious to assenting to certain propositions is fatal, and I agree with him.

Did I grow up with a religious background? Absolutely. Does that mean I believe in God? The answer is probably no. What matters to me is to get deeper than those questions, not to assent to them or deny them. Soon after I got to college I stopped going to church entirely. I found it untenable for lots of reasons, but I never stopped struggling with metaphysical and theological questions. What is the nature of being? What are the grounds of ethics? Those things have always mattered to me enormously. How to anchor one's life, how to know what's worth committing to. When I was seventeen I remember asking my father what he thought it was to be religious. His answer: "To find the world significant."

GS: How, then, would you describe your religious practice?

BCS: I don't use the word "religious" much. When I told my Dad I was quitting going to church, because I didn't believe the things that they were requiring me to affirm, he said I was probably right not to believe them. "The sad thing," he said, "is that you and your friends, if you opt out of religious community, are going to lose vocabulary in which to talk among yourselves about the things that matter to you most." Thirty years on I can report that he was largely right. It's extremely difficult to find words that come anywhere close to communicating, with most people I know, what it is that matters most to me. I find myself using different words with different people.

GS: Assuming, for a moment, your father's definition of leading a religious life, do you think that there are people who *don't* find significance in the world? Can you be a human being and not find significance in the world?

BCS: You can fail to find significance – but you end up not being fully human.

More seriously, I think that there's enormous dissatisfaction with respect to that question these days – people feeling that their lives are hollow or unsatisfying, people feeling anonymous, people feeling that their social and economic conditions don't give them a chance at a satisfying life, don't welcome them, don't provide them a way to participate, and so on. Many people, I think, don't find, or anyway don't find it easy to find, a deep sense of significance. And a large number of them feel frustrated and empty, as a result.

Think about the rise of religious fundamentalism in this country and in the Near East. You have the Christian right in this country, ultra right-wing Zionists in Israel, fundamentalist Muslims in the Near East and North Africa. I think these movements are all very similar. There is a deep unsatisfied hunger in many people's lives – an unfulfilled yearning, where people feel that materialist values, certain kinds of economic values, are not satisfying.

The fundamentalist movements recognize this lack, and are trying to provide answers. The problem is, I find their answers appalling. Without necessarily trying to, they cater to close-mindedness, bigotry and fascism. But what are we intellectuals and academics providing by way of response to that hunger, to that palpable yearning? What would it be for us to formulate a *better* answer – an answer that does justice to people, in their plural ways of being, an answer that does not have all of the bad aspects of fundamentalism that I worry about, an answer that is inspiring, in the literal sense of giving people hope, an answer that answers that palpable hunger for anchoring, for meaning, for significance? That's what we need. That, approximately, is what I want to do for the next twenty-five years: I want to help work on formulating an answer to that question.

GS: I wonder if the kind of anchoredness and significance you talk about can actually be found in science?

BCS: Well, it depends on what you mean by "science."

One of the things that people in science have historically tried to do, of course, is to subtract the issue of value. But we don't eliminate truth from science, which is a big value. If I come up with a theory that's false, that's not good! Even the most traditional scientist, therefore, has to agree that some norm is operating in science; namely, the norm of truth.

Given that, take the Greek separation of values into truth, beauty and goodness, and ask why science has hung on to the ideal of truth, and let go of the ideals of beauty and goodness. It is something of a default *modus operandi* for science these days to valorize truth, subtract goodness, and perhaps allow a little beauty back in to dance over the elegance of the equations. I want to argue for letting certain kinds of other values back in (especially ethical ones). I just want to do so in a way that respects why science originally threw them out, without compromising any of the truth to which science has been dedicated.

To see how this might go, think about what's happening with science, as we end the millennium. A while ago, I used the word "significance." For several hundred years we've had physical sciences. For several hundred years, that is, "natural" has meant "physical." What we are now going to have, maybe for another couple of hundred years, is a new kind of "natural" science, to go alongside the old one: something like semiotic or intentional science.

The original alchemists tried to turn iron into gold. I think of the world's C++ and Java programmers as "semiotic alchemists," trying to turn C++ code into gold. By now we have fifty years of a very widespread, inarticulate, absolutely dedicated and rather disheveled practice of people trying to construct arbitrary things out of symbols and information. It is a very similar situation.

So what does this have to do with religion? Signs, signifying, signification and ... significance! As we saw at the beginning, "significance" means *importance*. What is significant isn't just what has been mentioned or symbolized or represented or referred to, but what *matters*. That brings us to the million dollar question: is this new era of science – this new form of science of the "natural" – going to require a broadening of our sights to include not just factual, but also ethical values?

GS: Can you subtract the value of significance in your scientific study of signifying?

BCS: Some people would say that this is all a pun – that it is only an etymological accident that "significance," in English, means "importance," and is also used, more technically, to refer to the property of signs, whereby they signify things. But I deeply believe that's false. It's not an accident at all. Moreover, signifying is not the only thing that involves an ethical dimension, without people realizing it. Another notion that, surprisingly, is intrinsically ethical is the notion of materiality. Material evidence, in a court of law, isn't evidence that weighs some number of kilos, or that has an inertial mass, but evidence that makes a difference. Material objects, similarly, aren't just patches of the world that are made of physical stuff, but patches of the world that *matter* to us.

GS: The philosopher Ludwig Wittgenstein said that when science has answered everything it can, the questions of life remain untouched. You seem to be suggesting that the emerging science of semiotics may begin to say some things that do touch the "questions of life."

BCS: I disagree with Wittgenstein. I don't believe we can get away with saying, "No, science can't touch what matters. You have to look else-where." Statements like that, and Wittgenstein's claim, are both rooted in a particular conception of science – the one we've had for 300 years – which may not last. If I am right that a new metatheoretic frame-work is going to be needed to understand this new "Age of Significance," then maybe science will change to incorporate these new values. Stranger things have happened.

Just as science may change, so too religion may change, may *have* to change, into something unlike anything we have ever imagined. We may need a new theology unlike that of any religious tradition we've ever had, altered so as to capture the imaginations and inspire a world-wide community of diverse people, which can not only subsume all of science, but also incorporate the full range of human questioning into ultimate significance, give people a reason to live and an anchor for their commitments, help people understand why they care about the people they care about, why they should care about things that are important.

I don't claim to know what will fire the imagination, calm the spirit, do justice to the world, and provide grounding for our lives. All I know is that it is urgent that we start figuring it out.

Brian Cantwell Smith

God, approximately[1]

Preface

Slowly, but inexorably, science is encroaching on territory traditionally held
to be religious. Scientific accounts now deal with such topics as what it is to
be a person, the origins and character of altruism, sexual preference, marital
fidelity, first-person perspectives and the nature of consciousness. If understood as expansionist moves of a traditionally conceived science, these
developments may seem – especially to religious people – to reduce hallowed
categories to soulless physical arrangements, and thereby to desecrate the
sacred.

In point of fact, however, something much more important is happening
in intellectual history. These emerging developments reflect, and presage, a
profound shift in the foundations of science – as momentous as anything
that occurred in the sixteenth and seventeenth centuries (when natural
philosophy and theology originally sundered). Once the magnitude of these
changes is comprehended, the threat to what matters eases, and the way is
opened to an urgent intellectual task: reconciling our scientific, intellectual
and religious worldviews.

(*A note, in advance, to those who are allergic to religious language. You have my
sympathy. In fact you almost have my company. But I have come to believe that
questions that have traditionally been asked in theological contexts – about ultimate significance, what it is to be a person, the grounds of justice and humility,
etc. – are too important to leave hostage to the pro- and anti-religious debate.
I don't care, in the end, who professes religious affiliation, and who forswears it.
What matters is whether – and how – we answer those ultimate questions.*)

Introduction

Consider the rise of the religious right: the Christian "moral majority,"
ultra right-wing Zionism, Hindu and Islamic fundamentalism. These
extremist movements are responding to – and exploiting – a widespread social
dissatisfaction: a feeling that the reigning secular, scientific, economic and
political worldviews have failed to provide a sense of what matters: ways to

tell right from wrong, guidance to anchor individual lives, the wherewithal to resolve thorny ethical dilemmas.

I find many of the fundamentalists' answers appalling: bigoted, mean-spirited, scary. But what are the rest of us – we on the left, we scientists, we intellectuals, we in the academy[2] – doing about this heartfelt lack? If we don't respond to the yearning – if, willfully or unwittingly, we remain blind to the unsatisfied hunger – then we have no leg to stand on, in criticizing others' responses.

What we need are *better* answers: frameworks to stir compassion, give meaning to lives, combat prejudice, secure a modicum of economic well-being, preserve the planet. These frameworks must be global; it is too late for parochial sectarianism. And they must build on the best in science. We need to move forward, not back.

Now you might take this as a call to arms: that scientists (and I come at these issues as a scientist) should set aside their professional concerns, and take a stand. You might especially think this if you subscribe to what I will call the *classic dialectic*: the supposition that we can divide our understanding of the world into two parts:

1   A roughly causal, deterministic, value-free, third-person theory of *material objects* – a knowable, empirical science of the physical world; and
2   A more mysterious, phenomenological or experiential, inexorably first-person, value-laden *theory of spirit*.

A realm for the body, and a realm for the soul. It is a convenient dualism, an historically entrenched dualism – and still, unfortunately, a widely accepted dualism.

That it is still entrenched is betrayed by how many current views can be understood in its terms. Most obvious are the extremists: those who trumpet the idea that science is not only right, but comprehensive – and religion not only wrong, but useless; and (at the other extreme) religious ideologues, including some Creationists, who, hitchhiking a ride on postmodern critical theory, remind us that science is "just a story" – and then invoke divine authority, or *a priori* intuition, to assert the infallibility of received religious myth. But fundamentalists, of whatever stripe, are not the only classicists. Just as troubling, in my view – and equally in the thrall of the dialectic – are "quietists" who take scientific and religious perspectives as given, incommensurable, equally-valid perspectives on a common underlying reality. Neither view challenges the other, the quietists say – so they can be jointly and amicably held (again, in impeccable pluralist fashion). It is a convenient stance: do science during the week, go to services on the Sabbath – and laud science as showing the wonder of God's creation.

I don't believe any of these stances will do – extreme or accommodationist. The problem, I will argue, is much harder than that. The only tenable alternative – metaphysically, philosophically, theologically – is to reject the original dialectic and start over.

It is not hard to see why the dualism might need to be replaced. For one thing – something evident to all of us, since childhood – the world, *au fond*, is surely *one*, not two. That is not to deny what is right about cultural differences, to "disappear" the peculiarities of contingent experience, or to embrace the possibility of (imperialist) master narratives. There is no reason to, and many reasons not to, yearn for a single story. But it is cowardly to hide the task of reconciling our ultimate worldviews behind a convenient veil of pluralism. *How should we treat people? What is it, in the end, to be a person? What matters – to us, to others?* Do we really think we can inspire lives with incommensurable answers to such questions?

A second reason to challenge the dialectic is the simple fact that all the world's major religious traditions (not just the Abrahamic) are based on natural philosophies hewn before the rise of science. As many have noted, views of women, procreation, cosmology and a myriad other subjects are sadly in need of being brought up to date. It is nostalgic, at best, to assume that the profound changes in worldview the human race has undergone in the last few centuries could leave a religious myth intact.

Third – and not incidentally – shrinking distances and increased population in the global village increase the urgency of reconciling religious difference, not only between and among the major traditions, but between religious and secular worldviews.

Here, though, I will focus on a fourth reason to reject the classic dualism. *Modern science is rapidly encroaching on traditionally religious territory.* People have traditionally viewed science as dealing with the "merely physical," as is evident in the way the dialectic is framed; but it has been decades since anything so limited was true. Neuroscience is ablaze with excitement about elucidating the phenomenology of consciousness; psychology is pushing headlong into the realm of emotions, including jealousy, love and hate; gender differentiation and sexual orientation are staple subjects in biology; mathematical models are being developed of such basic notions as order, autonomy and self; evolutionary accounts are offered of altruism, tribalism and belief; computational models traffic in meaning and interpretation; first- and second-person points of view are being incorporated (along with third-person) in sciences as diverse as medicine, physics and computer science. Not since the nineteenth century has "natural science" been restricted to the physical.

Some find such developments sacrilegious. And, although I was trained as a scientist, I have some sympathy for such reactions. Sure enough, I'm willing to hazard, these developments *would be* sacrilegious – may even *be* sacrilegious – if understood in traditional terms. That is: if one understands these new forays and encroachments of science in terms of *old foundations* – if one attempts to layer these modern developments over a classical understanding of what natural science is – one is likely to conclude that we are doing exactly what religious people fear: developing a reductionist, material-istic picture of spirit, values, persons and other entities traditionally held to be transcendent.

But to understand current intellectual progress traditionally, like that, is to put new wine into old wineskins. And it leads to ironic results.

On the surface, this traditional approach – incorporating new scientific results within a traditional conception of science, without revising the core conception of what science is – seems like a perfectly scientific agenda. Indeed, something like it is probably the explicit or implicit goal of the scientific fundamentalists mentioned above. But it is a strategy doomed to failure. For one thing, for purely intellectual (science-internal) reasons, as I will argue, it can't keep up with the very scientific progress it is designed to celebrate. Furthermore – and more seriously – to defend the "cause of science" in this physicalistic or reductionist way, while allowing the debate to remain agonistically framed in terms of the classic dialectic, will play straight into the hands of fundamentalists. I worry, in fact, that it will lead to more of what we are already seeing, across the world's stage: the battle falling to the religious right. It is they, after all, who, as the issues are currently framed, have the premium on protecting a sense of significance, personal values, things that matter.

What is needed is something radically different. We need to develop *new foundations*, capable of meeting two simultaneous demands. We must:

1   reconcile the classic dialectic, by doing simultaneous justice to (i) what has traditionally been within the purview of science, and (ii) what matters most in the religious traditions – i.e., what lies behind the explicit belief structures of the religious traditions; and
2   update our foundational understanding of science, so as to accommodate new theoretical and technological developments, rather than assuming they will fit into a nineteenth-century conception of the "merely physical."

In the end, I don't believe that these mandates are distinct. Not only can they be simultaneously met; at the deepest level they come to the same thing.

In this essay, though, I will approach the issue asymmetrically, focusing primarily on the second task: updating our foundational conception of science. The main effort will involve taking stock of where we are now, examining the consequences of a variety of current results for the epistemic project of scientific inquiry, and making room for several impending new developments. At the end, I will suggest that accomplishing this "updating" task – i.e., formulating an appropriately reconceived foundation for science – will put us well on the way toward meeting the former goal of reconciling the intellectual and the religious.[3]

Turn, then – without preconception[4] – to the present state of science. As I will attempt to show, the tradition we are inheriting at the dawn of the new millennium is radically unlike the classical image – the conception inscribed in the traditional dialectic. In fact modern science is more unlike that conception, I believe, than anyone has yet recognized.

Twentieth-century science

Three twentieth-century scientific developments are particularly important.

First, what's gripped the imagination of the mythical person-on-the-street are results in physics: quantum mechanics and relativity, in the first part of the century – chaos, strings and fractals, more recently. Extremely odd stuff, especially at the quantum level: ten- or eleven-dimensional curved (maybe even branching) universes of strange forces governed by inscrutable logic. Even on a classical picture, modern physics is alien: a stupefying spray of interpenetrating waves of every conceivable frequency – turbulence, attractors, vortices smashing and piling up on top of each other in dizzying disarray. Imagine falling overboard in a storm – and opening your eyes to nothing but salt and spray. Now subtract you. It's a little like that, out there; only a zillion times worse.

A far cry from tables and chairs, obedient street lights, moribund committees, the PTA. So let's label this distinction. By the *physical world* I will mean the almost incomprehensibly strange, object-less world of modern physics. By the *material world* I will mean our familiar day-to-day realm of medium-sized macroscopic objects: people, cars, elections (and possibly even such arguably concrete entities as fame and détente). It is possible that the material world is, or can be, derived from the physical world by abstraction – but "abstraction" is a concept in epistemology, not in physics, and so itself stands in need of explanation.

So that's the first thing about twentieth-century science: the physical world is surpassingly strange, and quite unlike the material world. The second development – also relatively well known – is that scientific knowledge has proved *epistemologically recalcitrant* – far more so than was classically expected. Traditionally, a sharp line was taken to divide *knower* from *known*. But that simplistic model has turned out to be unsustainable in principle as well as practice. Relativity brought perspective into scientific claim; quantum mechanics shattered the myth that measurement was innocent; experiments proved more invasive than simple disquisitions on the empirical method suggested; debates rage about whether mathematical entities are part of the subject matter or mere theoretical equipment; considerations of (epistemic) tractability permeate scientific models. In human affairs, of course, many of these facts – the perspectival aspect of knowing, and the violent character of finding out – are almost truisms. What's novel is their implication deep in science: in the measurement problem, the collapse of the wave function, quantum indeterminacy, complexity bounds. Maybe all knowledge is violent, perspectival, implicate.

The third development I want to draw attention to (besides ontological and epistemological issues in physics) is the rise of the "intentional sciences": those dealing with symbols, meaning, reference, interpretation, truth. Logic has been under investigation for millennia, of course, but for most of that time it was viewed as ancillary *equipment for the doing of science*, rather than as itself subject to scientific (especially empirical) investigation. Starting in the nineteenth century, however, that began to change with Babbage's engines, Boole's *Laws of Thought*, the ground-breaking philosophies of Frege and Peirce. Their achievements flowered in the twentieth century, unleashing modern logic, meta-mathematics, psychology, linguistics, cognitive science – and the entire computer revolution.

Two facts about intentional phenomena need to be highlighted: a way in which they are, and a way in which they are not, "merely physical."

The first important characteristic of intentionality – the *non*-physical aspect – is most obvious in the case of reference. Something that is so self-evident, in our experience, as to defy our imagining its not being true, yet from the point of view of physical science seems almost miraculous, is our ability – with a simple few words, or the merest thought – to refer to things that are far away in time, space or possibility. I can refer to the Pharaohs of Egypt, without violating backwards causality; anticipate the great day on which the first woman will be inaugurated as U.S. President – without violating forward causality. I can refer to events outside my "light cone,"[5] with which physics prohibits any causal interaction at all. Moreover, this "arrow of directedness" is exquisitely precise: I can refer to the paint flaking off the ceiling exactly two centimeters in from the easternmost corner of the reception room in Mother Teresa's clinic in Calcutta, and my reference will "land" precisely there, with unswerving accuracy. Reference is zippy, too – traveling "at the speed of logic," as I once heard Alonzo Church say, we can describe the temperature on the surface of the sun without our reference taking eight minutes to get there. We can even refer to what doesn't exist: to the different situation we would be in if American presidents were elected by simple popular majority. And so on. Forget angels: reference goes where fools merely *imagine* treading.

There is a sense, that is, in which reference is physically transcendent. The amazing thing is that it manages to do these things without requiring spooky metaphysics. That is – and this is the second aspect of intentionality I want to highlight – there is also a sense in which reference is physically immanent. Referring outside your light cone doesn't contradict natural law. Somehow, without violating the inexorable spatial and temporal locality of physical law,[6] we are able to direct our thoughts outside the confines of the $1/r^2$ envelope to which physics restricts physical engagement.

Figuring out how we do this, I believe, is as important a twentieth-century discovery as any in science. It is not identified with any single name, though the giants of the early logical tradition – Frege, Russell, Gödel, Turing, Carnap and others – deserve a lion's share of credit. Unfortunately, their solution was codified within the formal tradition, and so is not publicly appreciated. ("Formality," it turns out, is a perversely abstract form of digitality: an assumption that useful theoretical categories can be completely, unambiguously, and absolutely divided – a particularly stubborn form of thinking in black and white.) But the insight can be extracted from the clutches of the formal tradition, to survive another day – or century.

Here is the basic idea. Physical regularities – causes and effects – are, as I've said, *local* in essentially all relevant respects (spatially, in the sense of three-dimensional proximity, and temporally, in the sense of temporal immediacy). That poses a problem for cognitive or intelligent creatures. All you *get*, if you are physically embodied – i.e., in terms of effective resources – is what is pressing in on you, right now, at the surface. You live in a laminar cocoon, that is, with physical coupling limited to the immediate here and now. Moreover, the world is *sloppy* (only weakly correlated), so you can't necessarily *tell*, from what is happening right near you, what is going on elsewhere – behind that rock, or back at home, let alone what went on yesterday, or will go on tomorrow. Fortunately, however, that same sloppiness – the local degrees of freedom – also means that an agent can rearrange its internal states with remarkable facility (if it's clever), without expending much energy. So what we do – what agents do, what it is to think – is to *represent the world out there*, beyond the periphery, by rearranging our internal configuration, and adopting appropriate habits and practices, so as to behave appropriately with respect to – develop hypotheses concerning, stand in appropriate relation to – that to which we are not, at the moment, physically coupled. What we can do, sometimes, is to exploit correlations between the incident, proximal effective array and the (non-effective) distal situations we care about (correlations called "information") – though the connection between the two is often intricate, involving lots of internal machination (called "inference").

It's not just *amazing* that semantics outstrips causal bounds, in other words, in the referential sense we spoke of a moment ago; that's what semantics is *for*. A "purely" physical entity – a patch of the world that hasn't figured out how to organize itself, locally, so as to be oriented toward what is distal – is existentially limited to the incident press of the immediate physical surround. Living in the here and now is cheap; that's what brute physicality gives you. Living in the there and then – that takes smarts.

Studying these things – how critters and computers can represent and reason and be directed toward what is distal, how they can think and act locally while honoring what is global – this has been the work of a 100 or so years of intentional science. And the fundamental principle on which it all works – the same principle that underwrites all of computing – truly is a "Big Idea."

So those are the three features of the last 100 years of science that I want to highlight: (i) the ontological weirdness of modern physics, (ii) the epistemological recalcitrance of physics in particular (and science in general), and (iii) the rise of what I am calling the "intentional sciences": logic, psychology, computer science, linguistics and mathematics (underwritten by a dash of philosophy). As we'll see, the three developments are related. They also have an interesting implication. On the classical image (figure 14.1a), *knower* was viewed as external to *known*. Physics has been epistemologically problematic in part because, as I said, physicists (and their practices) are implicated in physical knowledge, but physics is not a science that can explain knowing or practice. So physics has lived with an unresolvable predicament: though it aims for completeness, it cannot explain itself. It takes the intentional sciences to complete the picture, and resolve the tension (figure 14.1b). Since scientific knowing is one kind of knowing, and knowing is intentional, extending science to include intentionality brings *doing* science *inside* science.

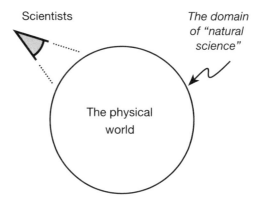

Figure 14.1a   Traditional "natural" (i.e. "physical") science

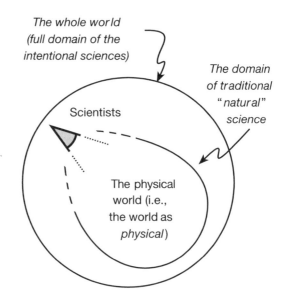

Figure 14.1b   The intentional sciences

State of the intentional art

The ontological oddity of physics, described earlier, is a substantive result. The other two developments – the recalcitrance of epistemology, and the reconfigurations mandated by the rise of the intentional sciences – are more structural. We've put intentional phenomena onto center stage, and redrawn the boundaries of inquiry; but we haven't yet said much about what intentionality is *like*.

In this section I want to dig deeper into this third development: the scientific exploration of the intentional. Only with some more specific results will we be able to see the full power of what is happening, in intellectual history, and assess prospects for ultimate reconciliation.

Once again, there are three things to be said.

From formality to participation

First, at the beginning of the new century, we are seeing something of a sea change in our understanding of symbols, meaning, thinking, interpretation and the like. The models of reasoning and knowing that were dominant in the first decades of the previous century, and that reigned in cognitive science and artificial intelligence (AI) as late as the 1970s, betrayed what we might call a *rationalist view from nowhere* (Nagel 1989). Intelligence was taken to be a process of (inductive or deductive) deliberation; agents were treated (at the relevant theoretical level) as abstract and unlocated; and the task domains that agents reasoned about were assumed to be ontologically unproblematic: neat realms of distinct, well-behaved, durable objects, exemplifying properties and standing in clear structural relations (rather as logicians imagine, when doing model theory).

Recently, however, this whole cluster of assumptions is being set aside. The distinction between agent and world is recognized as problematic, perhaps even illusory. Far from being abstract, agents are increasingly seen as crucially active, embodied, participatory – made of the same stuff as the domains in which they act. Perception, action and behavior are viewed as paradigmatic intelligent activities, rather than hypothetico-deductive ratiocination. *Driving around*, not *proving theorems*, is taken as emblematic.[7] Finally, it is increasingly agreed – especially in practical trenches – that those neat ontological assumptions about the *a priori* structure of the world don't work. The world simply doesn't come all chopped up into nice neat categories, to be selected among by peripatetic critters – as if objects were potted plants in God's nursery, with the categories conveniently inscribed on white plastic labels. A glance at the scan lines emerging from a robot's TV camera quickly dispels any such myth. On the contrary, a growing cadre of researchers believe that figuring out how to parse or "carve" the environment into workable pieces or "objects," in ways appropriate to the task at hand, is *the* major task that cognitive creatures face.

All of these changes – captured in such slogans as *situated* AI, *embodied cognition*, etc. – can be viewed as retreats from the aforementioned formal tradition: in various ways, they share a recognition that many classic dichotomies (inside–outside, symbol–referent, abstract–concrete, continuous–discrete) are *partial*: negotiated, plastic, problematic. Discovering and stabilizing such distinctions, when necessary, in project-specific ways, is an achievement of the subject (robot, animal, person), and therefore cannot be assumed by the theorist as *a priori* or given. Fitting an appropriate conceptual scheme to the world may lie closer to the heart of intelligence than working within one that is already established.

So that's the first result regarding the state of the intentional art: intelligence is concrete, messy and participatory.

Computability

The second result has to do with limits. Gödel's incompleteness theorems,[8] proved in the 1930s, rocked the mathematical and philosophical community. For the first time, it seemed, intrinsic limits were placed on possible intellectual achievement. But again, as with the "Big Idea" underwriting intentionality, the formulation of these insights in formal terms blocked appreciation in the wider intellectual arena of their true significance. One way to put the result is that *semantics* can never be wholly reduced to *syntax*. Pragmatically, though, in terms of the story told above, there's a simpler way to say it: although local, effective, physical arrangements can do a good job of standing in for remote, non-effective or abstract situations (especially in highly constrained circumstances) they are never *perfect*. In all but the most trivial cases, proximal surrogates can never entirely capture what matters about distal subject matters.

This moral was driven home by the "complexity" results of the second half of the twentieth century, having to do with *how hard*, in terms of space and time, it is to achieve things that are theoretically possible. It turns out that these "relative computability" results are much more consequential for embodied cognition than the absolute results discovered earlier. It doesn't matter that chess is finite when the number of possible moves in a game is on the order of 10^{120}. Remember: the number of possible 16×16 bit cursors on your computer screen is already 10^{78}, 10^{42} times smaller than the number of possible chess games, but still almost a million times greater than the total number of electrons in the universe. Combinatorics can kill.

Perhaps the most important consequence of these computability and complexity limits – especially when taken together with the weird ontological claims of physics – has been to drive an *irrevocable wedge between determinism and predictability*. Classically, it was assumed – most people still assume, I suspect – that if you *know* everything about a situation at time t, and that situation is deterministic (in the sense that what happens at moment $t+1$ is entirely determined by what is true at time t), then an intelligence – at least in principle, if not in practice – could figure out what the situation will be, at time $t+1$. *But it is not so* – at least not for any finite, concrete, embodied intelligence. In all but vanishingly few cases, waiting for the results to happen may be the metaphysically optimal – perhaps the only – way to know what will happen in detail (especially on the wonderful suggestion that the universe is running an optimal algorithm).

In sum: the impossibility of accurate (epistemological) prediction is entailed by quantum indeterminacy, computability limits, complexity results, turbulence and chaos and other aspects of non-linear dynamics, emergent properties and emergent objects ... on and on, through an astounding number of the major results of twentieth-century science. Though not the sort of thing that can be captured in a simple formula or theorem, there may by now be no more thoroughly established result in all of science.

Taken together, these first two features of intentionality (finite, embodied participatory creatures, subject to massively strong computability limits) radically undermine the classical image of "man" as a rational, all-knowing, *Übermensch*. In its place we get an ever-deepening sense of a world of paltry, finite, embodied creatures, struggling to make their way around in – struggling to make sense of – the world around them, using intrinsically partial, flawed, perspectival, incomplete, knowledge and skill. These aren't just practical limitations, either. Embodiment is necessary for reference, but it intrinsically (and radically) limits epistemic achievement.

It is a humbling image.

Dynamical norms

The third result of the intentional sciences is less familiar,[9] but if anything more consequential. It has to do with *norms*. I haven't said anything about norms, but to enter the realm of representation – description, language, interpretation, truth, etc. – is to enter a world of phenomena governed by asymmetric, paired evaluative predicates: true vs. false, good vs. bad, working vs. broken, beautiful vs. ugly, etc. – where one option is *better, more virtuous, more worthy*, than the other. Accurate descriptions are better than inaccurate ones; information is better than misinformation, helpful behavior is better than unhelpful behavior ... and so on. In fact one good definition of intentional systems is that they are just those systems that are subject to norms.

Truth is a famous norm – but not particularly general. Scientifically, moreover, it has been treated as "statical," in the sense of applying to (passive) *sentences* or *thoughts* – i.e., to *states*.[10] Full-blooded intentional systems, however, are *dynamic*, and hence governed by "dynamical" norms – norms that govern *process*. In logic (the seminal intentional science), the only dynamical norm that has received much attention is ontologically *derivative*, defined in terms of a statical norm. In particular, logic's dynamic processes (reasoning, deduction, inference to the best explanation[11]) are mandated to *preserve* or to *produce* truth, where it's assumed that truth and "best explanation" can be defined independently of, and prior to, their preservation or production.

This explanatory strategy – of starting with (presumptively autonomous) statical norms, and then defining dynamical norms in terms of them – has been adopted by other intentional sciences. Economic models of rationality and decision-making, for example, use dynamical norms of *utility maximization*[12] – where utility is (again) presumed to be static, explanatorily prior and autonomous. But the strategy doesn't generalize. And no computer scientist believes it. What computational experience teaches us is that things generally work in the opposite direction: the semantic content (meaning) of a symbol or expression or data structure is typically determined by (even: exists in virtue of) *how it is used* – i.e., by the role it plays in the overall system of which it is a part. Rather than define dynamical norms in terms of statical ones, that is, programers *derive statical norms from dynamical ones* – in an (often unwitting) endorsement of the Wittgensteinian maxim that "meaning is use."

If we get our statical norms derivatively from our dynamical ones, where do we get the original dynamical norms? What are they like? What governs, what puts value on, what evaluates, the use – i.e., the life and times, the activity – of general intentional processes? This question isn't usually asked so baldly, though a variety of alternatives are being explored. But the dynamical norm that is currently receiving by far the most scientific attention – in cognitive science, AI, evolutionary epistemology, research on autonomous agents, and of course biology – is evolutionary survival.

It's clear how you get a dynamical norm out of survival: a process or activity is deemed *good* to the extent that it is *adaptive* – i.e., to the extent that it aids, or leads to, the long-term survival of the creatures that embody or perform it. This idea of resting normativity on evolution has proved seductive. It has been used to define a notion of *proper function*, for example, in terms of which to decide whether a system is *working properly* or is *broken*. Thus the *function* of the heart is to pump blood, it's claimed, and not to make a "lub-dub" sound, because hearts were evolutionarily selected for their capacity to pump blood, not for their sound-making capabilities. Similarly, the function of sperm is to fertilize eggs because that's why sperm have survived (even if only a tiny fraction of them ever serve this function).

Most interesting for our purposes, however, is the use of this same idea to define semantic content (meaning, reference, representation, truth). The representation in the frog's eye *means* that a fly is passing by, some people claim, because it leads the frog to behave in an adaptive way toward that fly (namely: to stick its tongue out and eat it) in a way that contributes to the frog's (not the fly's) evolutionary success. Similarly, the shadow on the ground conveys information *about* the hawk in the sky to a mouse just in case it plays an evolutionary adaptive role of counterfactually covarying with the presence of hawks in a way that allows mice to escape.

Have we reached the end of the line? Will evolutionary survival be a strong enough dynamical norm to explain all human norms: justice, altruism, authenticity, caring, freedom and the like? I sincerely doubt it. But in a way that's not the point. For what is at stake is not what will ultimately subserve all the norms we need in order to understand human activity, but *what the dynamical norms are, in terms of which we understand activity as human (even: as humane)*. And that, I hope, is obvious: dynamical norms on human activity govern *what it is to live* – live well, be committed, do good, strive for what is right. That is: *ethics*. And not just ethics, but whatever governs *whatever* you do: ethics, wonder, curiosity, eroticism, the pursuit of knowledge for its own sake ... and so on and so forth, without limit.

In sum, science's taking on full-fledged dynamical normativity (our third intentional result) is unimaginably consequential. It implies that any viable account of intentionality – any transformation of science broad enough to incorporate intentional systems, and thus to treat meaning along with matter and mechanism – will also, thereby, *have to address mattering as well*. Put it this way: in spite of what the logical tradition may have suggested, you can't just bite off *truth* and *reference*, and glue them, piecemeal, onto physical reality, without eventually taking on the full range of other norms: *ethics, worth, virtue, value, beauty, goodness*. By analogy, think of how computer science once thought it could borrow *time* from the physical world, without having to take on *space* and *energy*. It worked for a while, but soon people realized what should anyway have been predictable: that time is not ultimately an isolable fragment (not an "independent export") of physics. By the same token, it would be myopic to believe that the study of intentional systems can be restricted to some "safe" subset of the full ethical and aesthetic dimension of the human condition – and especially myopic to believe that it can traffic solely in terms of such statical notions as truth and reference, or limit itself to a hobbled set of dynamical norms (or even an isolated case, such as survival).

Moreover, to up the ante – in case this seems too mild – something else, if anything even more consequential, is implied by these same developments (and with this the pieces of the story all start to fit together). I said that the classical model of intentionality assumed that the meaning of symbols and representations could be assessed in terms of the objects and properties in the world that they corresponded to, independent of how those symbols and representations were used. But I've also said that few working scientists believe the classical model any more – in part because the physical world doesn't supply the requisite objects (remember, representational contents need *material* objects, which physics doesn't supply). That means that it is incumbent on a theory of representation to explain the objects that figure in the content of a creature's representational states. Objects, that is, are to be explained in terms of the normative structures governing the representations whose contents contain them. And those norms, we've just realized, are ultimately grounded on *dynamic activity*.

The material ontology of the world, in other words – what objects and properties there are, for a given creature (not just what objects and properties the creature *takes* there to be, but what objects and properties there *actually are*, in the world, for that creature) – will be a function of that creature's projects and practices. That is, as anthropologists, phenomenologists and postmodernists generally realize, you can't identify the objects first and then tell a story about the lives people live involving them. Rather, the objects themselves – what things exist, what type of thing a given entity is, what differentiates one thing from two – can only be determined with reference to the lives lived in their terms. For high-level social entities this isn't surprising: date-rape didn't exist, I take it, for aboriginal singers of Australian song-lines; baseball's strike zone isn't part of the furniture of the world, for earwigs. But the present claim is more radical: it says that what is the case for date-rape and strike zones is also the case for food, clothing, rivers, people – perhaps even for the number four.

Ontology is inextricably linked to epistemology, in other words, and epistemology inextricably linked to ethics. And not just onto*logy* and epistemo*logy*, in the sense of the *study* of things and of thinking. Things and thinking *themselves* are inexorably linked, and linked in turn to life and the good. Fair enough; these are conclusions we scientists can live with; they are also consonant with a thoroughgoing rejection of formality. What is striking about them in the present context, however, is that we have come to them by making two seemingly innocent moves: (i) by understanding the limited contribution physics makes to material ontology; and (ii) by recognizing that dynamical norms have explanatory priority over statical ones. That is: we have come to these conclusions not as meta-scientific attitudes – matters of preference or stance – but as *science-internal results*.

Not only has doing science been brought within science, so have norms, values and mattering. And not just values, but fundamental questions of ontology and metaphysics, too. For example, which of realism, irrealism, formalism or idealism, is right? Indeed, the answer to this question is staring us in the face. Metaphysically, the world is one. That's an anchor of scientific inquiry, to say nothing of common sense: no matter how disparate our cultures, your insecticides pollute my water supply; my car bomb rains on your parade (Smith 1996: 100). Ontologically, though, our worlds are many. They are many because objects and properties involve abstraction – and *abstraction is normatively governed*. Our objects are as constitutively different (and, of course, as constitutively the same) as our policies and practices.

We can summarize this conclusion etymologically. A *material* object is something that *matters*. It must matter, in order for the normative commitment to be in place for the objectifying creature to take it *as* an object: to be committed to it as a denizen of the world, to hold it responsible for being stable, obeying natural laws, and so forth – and to box it on the ears when it gets unruly. It's no pun, in other words, or historical accident, that we use "material" as a term for things that are concrete (made of "matter") and also as a term for things that are *important*, as in "material argument" or "material concern."

In fact that's one way to see where the intentional sciences are heading: they are going to have to *heal the temporary rift that for 300 years has torn matter and mattering apart.*

The age of significance

In a moment, it will be time to combine these results about intentionality with the general scientific developments rehearsed earlier in order to take stock of the present and future state of science, and to ask about the prospects for reconciliation. But one issue needs to be addressed first – to deflect misunderstanding.

If you ask most scientists – including modern intentional scientists – whether they *think* they are engaged in partially irrealist metaphysics, probing the ethical structure of the human condition, or constructing scientific models of the good life, their answer, it's safe to say, will be: *no.* But that is not the question at hand. It doesn't matter, for our purposes, what people (currently) *think* they are doing. As social critics, philosophers of science, and writers in science studies have repeatedly emphasized, scientists are not usually very reflexively self-critical, or necessarily experts about the nature of their own activities. Rather, what will matter in the long run (irrealist sentiments notwithstanding) is what they – i.e., what *we* – are *actually* doing.

That said, I will confess that, when considering the present situation, I am sometimes reminded of cartoon figures who run off the edges of cliffs and then hang there, motionless, for a moment, until they look down – and only then fall. A great many cognitive scientists, computer scientists, evolutionary biologists, linguists, logicians and the like, in my view, have already run off the edge of the "natural science" cliff, but haven't yet looked down. They have long since abandoned the allegedly "safe" *terra firma* of pure, local, causal explanation and pure, unadulterated physical phenomena – mainstays (even bastions) of science as we know it. Locally – in the thick of moment-to-moment interchange – they defend a variety of sensible views, such as what a data structure means depends on how it is used. What few seem to realize is how profoundly such seemingly innocent adjustments shake the foundations of what for 300 years we have called "natural science."

To make this concrete, it may help to consider the case of computing – one of the intentional sciences as already mentioned, and (as it happens) my own area of expertise. For many years I have been engaged in a foundational inquiry into the nature of computing – trying to figure out what it is, where it came from, what its intellectual importance is, what it augurs for the future. After thirty years, the project is largely complete ... and I have failed. Or rather: I have succeeded, I believe, in coming up with the answer. But the answer is: *there is nothing there.*

Here's the point: computers involve an interplay of the two things we have been talking about since the beginning: (i) *mechanism*: in the sense of a materially embodied, causally efficacious process, wholly grounded in a physicalist metaphysics; and (ii) *meaning*: in the sense of a realm of symbols, information, representation, norms, and the like. It is universally believed, however, that, in addition to these two things, computers are somehow *special*: that they involve some *particular* interplay of these two issues, have some distinctive or characteristic identity – digitality, for example, or formality, or abstractness – that makes them a worthy subject matter for scientific investigation. That computation is special is implicit in the idea that there might be a *theory* of computation.[13]

That is what I now claim we will never have. Admittedly, and somewhat distractingly, there does exist a body of work called the "theory of computation" – but I would go to court to show that it is *wrong*, in the following strict sense: *in spite of its name, it is not a theory of computing after all*. That is not to downgrade it. What the received (so-called) "theory of computation" actually *is*, I believe, is something of incalculable importance, worth a passel of Nobel prizes: neither more nor less than a mathematical theory of causality. But it isn't a theory of computing, because it deals with only one of computing's two constitutive aspects: with mechanism, but not with meaning.

The ultimate problem, however, is not with current theory; it is deeper than that. We will *never* have a theory of computing, I claim, because there is nothing there to have a theory of. *Computers aren't sufficiently special.*[14] They involve an interplay of meaning and mechanism – period. That's all there is to say. They're the *whole thing*, in other words. A computer is anything we can build that exemplifies that dialectical interplay.

While that might seem like a dismal result, in fact I believe the opposite: that it is the most positive result that a computational triumphalist could possibly hope for. Moreover, it is a result that has a direct bearing on our present subject matter. In fact – this is why this digression was warranted – we can lift up this "nothing special" result about computing, integrate it with the general results about intentionality surveyed above, merge those into long-term scientific developments, and finally get a fix on what exactly is happening to the foundations of intellectual inquiry as we enter the third millennium.

The situation is depicted in figure 14.2. Time runs along the bottom axis – from the fourteenth century up through the present and into the future. The vertical axis represents something like importance or weight. On the left is natural science, rising in the sixteenth and seventeenth centuries, peaking in the nineteenth and twentieth. To its left is an indication of alchemy: a rag-tag bunch of exploratory practices, conducted in a disheveled and a-theoretical way, that for several hundred years after Newton were shunned as "unscientific," but that are now recognized (i) to have involved far more knowledge and subtlety than for a long time was realized, and (ii) to have served as a crucial precursor to the very possibility of the emergence of an intellectually satisfying, theoretical mechanistic science.

What about computing and the intentional sciences? They are represented on the right. Whereas traditional natural science embraced causal explanations, and dealt with mechanism and matter – i.e., with purely physical stuff, in its myriad forms – the full range of intentional sciences, as I've said, involve a dialectical interplay of mechanism with issues of meaning, norms, interpretation, semantics, ontology and the like. That is: they deal with issues of *significance*. Moreover, what computer science (or computational practice) is, I believe – what history will ultimately recognize it to have been – is an experimental, synthetic precursor to the emergence of this new era: i.e., something I call "semiotic alchemy." Think of all those C++ and Java hackers, trying to turn Web pages into gold. And think, too, of the profusion of rag-tag practices conducted in cottages, basements and garages, that constitute "computing in the wild": they embody a vast wealth of pragmatic and practical understanding, albeit in pre-theoretic fashion.

Computers, that is, are *the laboratories of the intentional sciences*. They allow us to conduct experiments with language, meaning, interpretation, function, normativity, perspective, thinking – experiments of middling complexity, between the frictionless pucks and inclined planes of mechanics and the full-blooded complexity of the human condition. Computers give us raw material, in terms of which to experiment with – and thereby come to understand – the primordially intentional.

What this all means is that we are gradually but inexorably undergoing a social and intellectual transformation every bit as consequential as the Scientific Revolution. We are leaving the age of (purely mechanistic or physicalistic) science, and entering what I have called an "age of significance" – an age in which mattering, living truly (as much as speaking the truth), importance, etc., will take their rightful place in the intellectual pantheon alongside matter, materials and mechanism.

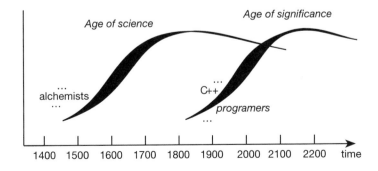

Figure 14.2 The age of significance

I am not sure whether to call the new age "science" – i.e., to assume that the term "science" will broaden to include the new kinds of understanding we are talking about – or whether "science" will retain its use for the kinds of causal explanation and physical phenomena legitimated in the past 300-year era, and something else will be introduced for the intentional variant. But given that "*scio*" is just Latin for "know," "natural" presumably means "whatever isn't supernatural," and thinking and referring (once one lets go of the hegemony of physicalist science) is about as natural a form of activity as there is – and with an eye, too, to the power and prestige that the scientific enterprise wields in society – I would guess that the term "science" will be extended to incorporate the new sense. However: just as I started with an apology to those allergic to the term "religion," I want to be first in line to say, to those allergic to the term "science," that the transformation in understanding we are on the verge of – if the diagnosis I am suggesting is right – is of almost unparalleled magnitude. This is not science as anyone has ever known it. It is something profoundly novel – something liable to change forever the reservations people have had (perhaps correctly) about the impossibility of explaining ultimate questions in scientific terms.

Reconciliation

We return, finally, to where we started. The call to arms, we can now see, is not that science should change its tune, and take on a radically new set of questions. Nor is it that scientists should set their scientific work aside, and devote one day a week to religion. The call to arms is scarier than that.

Whether it realizes it or not – whether we *want* it to or not – science is *already aiming in a direction* where it will take on questions of unprecedented weight. Not just traditionally scientific questions of great social or moral impact (such as whether to develop nuclear weapons), but questions of *what it is to be weighty, what it is to be serious – even: what should matter*. Traditional concerns of the religious traditions, that is, are being subsumed within the scientific juggernaut, independent of preference or protestation.

So the gauntlet – the call to arms – is simply this:

> *Will our scientific and intellectual answers be strong enough to live by, to sustain the world? Is our moral vision and political fiber tough enough – does our understanding go deep enough – for our answers to be able to counter the influence of the religious right? As scientists, academics, intellectuals – are we up to the task?*

I don't yet know how to answer that question. But answer it we must. And if what I have said here even points toward the truth, then maybe – just maybe – answer it we can. The possibility of our answering it positively depends not on the fact that science is domesticating territories traditionally considered religious, but on the deeper fact that science itself is in for an almost total metaphysical overhaul. If we can open our eyes to those transformations, and do a responsible job of stewarding it (and ourselves) through the upcoming changes, then – and only then, I believe – do we have a chance of succeeding at the project indicated at the beginning: of founding a vision strong enough to underwrite the world.

And remember: it is the *whole* world that is at stake: mattering, as well as matter – significance, as well as silicon.

For just a hint of how this will go, let us turn, one last time, to intentionality.

I said earlier that one characterization of intentional phenomena is that they are subject to norms. Another, widely attributed to Franz Brentano, was implicit in our discussion of reference and non-locality: the fact that intentionality involves *orientation* or *directedness*. We have already seen this in the case of semantics: to be "about" something is to be *oriented toward* it; to think about something is to cast your mind that way; to mention something involves a directed commitment. But orientation – a profound kind of directedness – is even more powerful than those examples suggest – more important, even, than Brentano may have realized. Rather than talk about it in the abstract, though, I want to illustrate it by considering one of the questions with which we started: what it is to be a person, what it is to be *us*.

To start with, we are here – not delineated as such, in physical force fields, but physically instantiated nevertheless. We are *in* and *of* the world; that much is a consequence even of traditional science. As intentional agents, moreover, we exploit the capacities of our physical embodiment to transcend the proximal, causal limits of that same embodiment. That is: we reach out, commit ourselves, interact, start taking the world *as* world. It is that inchoate directedness – that fact that we're not just *in* and *of*, but also *about* the world – that starts us on the long and difficult road toward humanity.

Rocks are in and of the world, too. But unlike us, they're not directed toward the world. They, like us, are infinite; and infinitely various – but they don't know it. In a way, there is nothing less transcendently splendid about a rock than about any of the rest of us; in that the Buddhists are right. But rocks are immune to their own transcendence. Nothing much *matters*, for rocks. That's in part because rocks aren't even *rocks*, for rocks. Not being abstracting, intentional creatures, they're incapable of distilling the world's richness into predicates like "rock," much less into individual entities like *a* rock.

We *can* take rocks to be rocks. Doing so requires commitment. Moreover – and this matters – it requires commitment not just to the rock, but also *to the world in which the rock exists*. In order to abstract anything as an object, that is, in order to construct material ontology, we have to be committed to that out of which, and that in which, we objectify. It follows that, in order to say anything at all – in order to refer, in order to stabilize an object as an object – an agent must literally be committed to that which cannot be "said." For the "world," in this foundational sense, is not the "post-intentional" world of thereby-arrayed material ontology. Ontology, after all, as we have already seen, is plural; the world is prior, unitary – and more profound. It's something like the "ground of being" that, in our faulty, partial, perspectival, self-interested way, we take to *host* the rock, or – if one likes Kantian language – something like the transcendental grounds for the possibility of objects. Or the world of "no-thing-ness," if one's preferences run Buddhist (remember: the gauntlet is to develop a global perspective). Perhaps it is a noumenal world, except that to cleave phenomenal appearance from noumenal reality sounds suspiciously like one of those formal distinctions we have been at such pains to eschew. Whatever it is, it is a world of norms as much as of objects, a world of mattering as well as of matter. Maybe it is a world of enchantment – maybe even a world we can re-enchant.

Put it simply. The world that science (without knowing it) is leading us to, or at least the world to which we are being led by whatever science is morphing into – the unutterable world, the ground of being – is far *less* like the mundane physical world classically set up in contradistinction to God, and far *more* like what the religious traditions, if I understand them, took God (or anyway, the ultimate realm of the sacred) to be. It is because of that fact, and only because of that fact, that we have a prayer of forging a sustaining intellectual vision.

Moreover, it is because our *commitment* is to this unutterable world that *orientation* is such a powerful notion. For notice how much directedness covers. Directedness underlies (and is prior to) purpose or *telos* – directedness in time. Directedness underlies (and is prior to) reference and truth – directedness to *what is the case*. Directedness underlies ethics: loving, fighting for justice, treating the world with kindness. And it underwrites curiosity, wonder – even reverence and awe.

It is this common application – to truth and compassion and justice and generosity and beauty and perhaps even grace – that makes Brentano's suggestion so poignant. All sorts of virtues – getting out of bed in the morning and lending a hand; shouldering responsibility for family and friends; taking things seriously; accepting responsibility for the consequences of one's actions; accepting responsibility for one's unutterable particularity; *accepting responsibility for the inevitable violence one does to the world by describing it at all* – all these things involve orientation and directedness: orientation up and out of oneself, to the encompassing world as a whole. Our orientation is not to the world as *other*, since each of us is inexorably part of that world (the world has no *other*); and not to the world as *object* – for the world is not one, in any sense in which it could have been two (and anyway objects are post-intentional, whereas the world is prior) – but to the world *simpliciter*: the world entire, the world of which we are a part – a world that so spectacularly defies description that the very notion of "description" is defined over and against it, as a way of watering it down. A world of matter and a world of mattering, a world in whose significance our own significance rests, a world unpredictable and risky and hard to master, a world to fight for and preserve, a world to struggle with, play in, defer to.

It may not be God. But it might be enough.

Acknowledgment

Thanks to: the members of the "Science and the Spiritual Quest I" workshop on information technology: Michael Arbib, Char Davies, Anne Foerst, Kevin Kelly, Mitchell Marcus, Mark Pesce, Henry Thompson and (the late) Mark Weiser; to Margaret Wertheim and Billy Grassie for their participation; and to Mark Richardson, Philip Clayton and Bob Russell, for organizing the conference. For comments on more recent drafts, thanks to John Coleman, Gillian Einstein, Will Oxtoby, Tom Settle and Arnold Smith.

Notes

1   This essay is dedicated to the memory of my father, Wilfred Cantwell Smith, July 21, 1916–February 7, 2000, who not only inspired my interest in these (and many other) topics, and taught us what it was to be intellectual, but read and commented on a draft of this essay, just a few months before he passed away. Although our perspectives are superficially very different, one doesn't have to dig deep to see the overwhelming debt that is owed to him. One of my last promises to him was that I would endeavor, as best I could, to follow in his tradition of wrestling with questions that matter.

2   This is not to suggest that these four groups are coincident.

3   I might say that the sort of reconciliation I discuss in this essay doesn't even begin to consider issues of practice: ritual, community, meditation, etc. Here I am only concerned with reconciling underlying worldviews. It will be noticed from what follows, though, that I don't ultimately believe that worldviews are separable from practice – so there is much work to be done.

4   Or anyway with as little as possible.

5   Physicists define the "light cone" of an arbitrary point to be that region of space-time with which communication or transport is possible – i.e., with which interaction can happen at less than or equal to the speed of light. If you imagine your current self to be at a point in four-dimensional spacetime, your light cone consists of two infinite cone-shaped regions: one, containing all events that could have affected you to date, converging in toward you, from the past; another, containing all events or points that you could possibly affect, from now on, spreading out away from you, into the future.

6   By "temporal locality" I mean immediacy: the fact that an event, at time t_1, cannot influence another event, at time t_2, without passing that influence through all intervening moments t_i between t_1 and t_2. Temporal "action at a distance" is as forbidden by the laws of physics as spatial action at a distance. There is this nagging issue: some non-local phenomena appear to be validated by modern quantum mechanics. In spite of some suggestive but extremely inchoate reasons to suppose that quantum non-locality may be essential for intentionality, macroscopic locality (spatial and temporal) remains the dominating "feature" of the world that intentional phenomena surreptitiously devise (without violating physical law) to referentially circumvent.

7   It is interesting that Descartes is being turned upside down. Whereas he thought that perception and navigational action were relatively simpler (accomplishable by "mere brutes"), rational high-level reasoning, he thought, was beyond the

powers of ordinary physical devices. The recent history of AI and cognitive science suggests that high-level symbol-manipulating logical inference (in at least some cases) may be much easier for machines to achieve than in-the-world perception and action.

8 Gödel proved that no formal (syntactic) axiomatization of mathematics could entirely capture what we intuitively (but correctly, most mathematicians think) take mathematics to be.

9 In fact I know of no explicit account of it; the formulation presented here is my own.

10 By "statical norms" I don't mean norms that don't change, over time. Evaluative metrics on book design, or abstract art, may evolve through the ages, but they would still be counted as statical, on this typology, because what they are evaluative predicates on – books or paintings – are static things. I use the (simpler) phrase "static norm" to denote norms that don't change – that are temporally fixed. By the same token, a *dynamical* norm is one that holds of or governs processes; a *dynamic* norm, one that itself changes, over time. Thus statical norms can be dynamic; dynamical norms, static. (The distinction is parallel to more familiar distinctions between "historic" and "historical," "strategic" and "strategical," etc.)

11 A norm, devised as a simple model of scientific theorizing, in which one aims to infer, from data or some other set of "facts," the best possible explanation of those facts.

12 I.e., a process of maximizing some utility: the worth of an investment, the pleasure of the participants, the good of society, etc.

13 It is also implicit in the idea that underlies AI and much of cognitive science: that we might be computers, too, in a way that isn't tautologous, isn't obvious.

14 That is, computers in general: not just present-day computers, but past computers, future computers, computers we will build and computers we will never build – computers in the fully general sense (which is what a theory of computing, if there is one, should be about).

References

Nagel, Thomas (1989) *The View from Nowhere*, Oxford: Oxford University Press.
Smith, Brian Cantwell (1996) *On the Origin of Objects*, Cambridge: MIT Press.

15 Arthur Peacocke

For over twenty-five years *Arthur Peacocke* taught and did research on biological macromolecules (especially DNA) in the Universities of Birmingham and Oxford, where he was a Fellow of St. Peter's College. In 1971 he took Anglican orders and in 1972 became Dean of Clare College, Cambridge, where he worked on the interaction between theology and science. He has been a Bampton and a Gifford Lecturer and until recently was Director of the Ian Ramsey Centre, Oxford, for the study of religion in relation to the sciences. He is Warden-Emeritus of the Society of Ordained Scientists (S.O.Sc.) and a Hon. Canon of Christ Church, Oxford. His recent books include: *From DNA to Dean: Reflections and Explorations of a Priest-Scientist* (Canterbury Press, 1996); *God and Science: A Quest for Christian Credibility* (SCM Press, 1996); and *Paths from Science towards God: the End of All Our Exploring* (Oneworld, 2001).

Interview by Philip Clayton

PC: Could you begin by saying a few words about your religious background?

AP: The Church of England is my sort of church. It's very open. It has a basic liturgy – in its older forms of great beauty of language – but it's practiced in many different ways. When I was a student at Oxford, William Temple, then Archbishop of Canterbury, recounted his experience in studying philosophy at Oxford. The last term as you come up to Finals is very crucial. His tutor advised "You should go back and read Plato again." He did this and then realized that all the major questions, all the things with which he had been concerned for the last three and a half years, had all been perceived and raised, if not solved, by Plato. The great questions were still the great questions. He went on to say "This is how the Gospels work in the history of Christian experience." We come back to them with new insights and find that the basic questions raised are all there. I came away from the sermon thinking, "Perhaps after all, Christianity is believable with some intellectual integrity."

One is a fool not to go back to the roots of a tradition just because you think you've outgrown it. From that point I gradually became aware that there was a real strand in the Church of England which has always maintained a liberal approach, namely that faith is based not only on tradition and the Scriptures, but also on reason founded in experience. Theology should have the same criteria of reasonableness as any other respectable intellectual exercise, namely those of coherence, fit with the data, consistency and fruitfulness. I think any intellectual quest in any discipline needs to value these, though the data with which different disciplines work can be very different.

PC: But do the criteria you've just spelled out apply to every Christian believer?

AP: I presume that nobody would become a Christian or engage in Christian activity if they didn't think it made some sort of sense for their lives and the world they know. Different people bring different intellectual resources to bear upon it.

PC: What are the parts of broader religious practice that stand outside of this requirement?

AP: The experience of love and caring for other people, of prayer, of worship, and of aesthetic and personal elements which are never going to be subsumed into scientific categories.

PC: But it's not necessary as a Christian religious believer to take anything on authority, as given?

AP: No, I don't think so. Because there's always the problem of: where is the warrant for these authorities? The biblical tradition is a tradition of going through your own experience and reshaping it in the light of experience. Religious tradition helps. It gives you resources of language in which to try and express your experiences. One of the impoverishments of modern life is that because people have dissociated themselves from major religious traditions and think they are starting with a *tabula rasa*, they have no vocabulary or resources to express their own personal experience. That's why a worshiping community is important, I think.

PC: So much of your work is based on establishing the credibility of theistic belief in light of modern science. Do you ever struggle with that tension?

AP: Oh, all the time. If you take part in the liturgy, as I do, you're bound to. The symbols and languages of the traditions handed down to you have had many different meanings poured into them through history. We are bound to be using these terms differently from people in the past, although we have the intention to refer to the same realities. Insofar as the words that are used sometimes fit uneasily with what we want to say, then perhaps we should be seeking new symbols. Meanwhile, because they have such rich associations in the liturgy and in devotion, one goes on using these terms and terminology. I use trinitarian language because I do think God comes to us in three distinctive modalities, transcendent, immanent and incarnate. I could spell that out in various ways, but how that relates to the actual ontological nature of God, I don't know. If God is more than a simplistic unity, there must be a diversity-in-unity in the Godhead, which will always be beyond our understanding.

PC: Wouldn't there be a strong disanalogy to the scientific case?

AP: One has to take into account the nature of the object one is talking about, namely the natural and biological world, which is the focus of science. This is more accessible, more available to public discourse than the experience of ultimate Reality, which we name as God. God is a *deus absconditus*. God doesn't publicly flaunt God's presence in a way which is publicly irrefutable.

Even to say your prayers or to worship, you've got to be convinced with integrity that the words you're using refer to some reality other than one's self, that you're using the words as accurately as you can. In the case of worship, this means that metaphor and imagery and poetry and symbolism play a much bigger part in trying to refer to reality as you experience it than they do in science. But even science has its symbols which you have to learn and understand. Mathematics is full of them. You have to learn to understand the symbolism of Christian worship. You can't, with integrity, do mathematics unless you think your symbols, your terms, stand for something. You can't worship unless you think the words are referring to something.

PC: Our next major area is this question of purpose and destiny, as treated in the physical sciences and in theology. I guess the question would be that Christian theology would seem to be better off in a period in which the physical universe seemed to manifest a purposive development, or an ordered development.

AP: Stones falling to the ground were said to be explained because they had a *telos* to do so. But it wasn't a purpose being worked out in time, in the physical and biological world. That only came along as a question when there were evolutionary perspectives. It was still a static world with given properties and the concept of *telos* was used as an explanation. Aristotle's final cause was used within that world. Of course, Newtonians showed that this wasn't a necessary way of interpreting the physical world at all. So it was only a qualified help.

PC: I hear you saying that although there is evidence from the study of the natural world for theism, in no period between Aristotelian science and complexification theories of the last fifteen years, could we speak of a stronger or a weaker defense of theism from the physical sciences.

AP: That's a complex historical question. I don't think I have enough knowledge of the history to say that, quite honestly. There have been individuals who have found the connection, and I suppose the growth of skepticism in the nineteenth century was actually more linked with political changes, and also with the interpretations of the role of chance, which I'll come to in a moment, in evolution. But up to then, I think people could be moderately comfortable with the sense of the ordered world, ordered by mathematical law, as being the expression of a divine Mind, and therefore consistent at least with Christian theism, and indeed with Islamic and Jewish theism.

PC: You understand what would make the skeptic uncomfortable here – that science has changed radically through two thousand years, but allegedly it has always been evidence for theism. The skeptic says if it's of that sort of generality, with such huge vacillations in scientific knowledge, it's not much of an argument for purpose.

AP: It's not an argument for purpose, but an argument for taking seriously the inherent rationality of the universe. Why should the universe be rational? Why should it be amenable to rational investigation and manifest rational properties? It was Kepler leading to Newton who began to see mathematical laws as being universally valid. Their belief that there is a source for the existence of the universe of this rational kind is a better explanation of the existence of the universe than that it just happens to be like that.

If you were designing a universe and you wanted these structures to have the capacity to change and develop, the only way you can do this is by introducing some randomizing element so that these structures would exist for a certain period and then dissolve into their units to be replaced by other structures which had new capacities. You could only do that by letting the structures exist only for a certain amount of time and be able to copy themselves, with some copies winning at the expense of those that have not got the modifications. In other words, you have natural selection.

If you're going to have a universe which has an in-built creativity capable of producing new structures and new forms, you end up with the inevitability of the short-lived existence of the entities, that is, death. If you are going to have a universe which has structures sensitive to their environment that can change, you're going to have nervous systems and therefore pain. If you're going to have structures which eventually develop sufficient awareness and information-processing so they can monitor themselves and have communication with other entities, you're going to have suffering. So if you're not going to have a clockwork universe which is totally static and totally unchanging, you're into something like natural selection, even "natural evil."

If you were God and wanted a universe to have the capacity to change and produce a particular kind of entity, but also to be in a law-like framework, you're going to have to introduce a randomizing element along with a law-like framework. You need the balance of the two to have a creative universe. This is where the purpose might come in. If you wanted a creature to have all these properties and to be able to respond freely in their own way, to circumstances, you're going to have something like our present universe.

Arthur Peacocke

The challenges and possibilities for Western monotheism

Introduction

The advent of a third millennium of the presence of Christianity in the world evokes variegated emotions in its followers. Looking back one could ask, with the apocryphal, and somewhat seedy, heckler of the Christian evangelist at London's open-air forum in Hyde Park: "Christianity has been in the world for nearly 2,000 years and look at the state of the world!" To which the prompt reply was "Soap has been in the world for 4,000 years and look at the state of your neck!" Minimally, like all religions, Christianity has been both the focus of idealistic aspirations and the milieu within which human fallibilities, and even wickedness, have been exercised. The recent arrival of the year AD 2000, "Of the Lord," cannot but provoke reflection on the whole sweep of Christian history and of the situation of that faith as we enter its third millennium.

Insofar as we focus on "Western" societies, the most prominent counter-culture to Christian faith has not been Marxism, which has risen and fallen, but a materialist secularism engendered, it is widely believed, by the dominance both of scientific ideas and of science-based technologies in our lives. The all-pervading effect of the latter I must leave to the social historians and social psychologists, but the former – the effect of scientific ideas – is certainly within the scope of the Science and the Spiritual Quest Project. Early on, in the first quarter of the last century of the second millennium of Christianity, the philosopher-mathematician Alfred North Whitehead could aver with prescience that

> the future course of history would depend on the decision of this generation as to the proper relations between science and religion – so powerful were the religious symbols through which men and women conferred meaning on their lives, and so powerful the scientific models through which they could manipulate their environment.
>
> (quoted in Brooke 1991: 1)

Now, as we approach the end of this century, we can see that the task he gave that earlier generation is still incomplete; only now is the study of the interaction of science with religious perceptions in general, and Christian ones in particular, being undertaken with any rigor and sophistication in our universities and becoming the concern of thinking believers. To what extent this interaction is proving challenging and fruitful, or merely destructive, I will advert to later. But first let us recall the various cultural milieux in which what we have known as the natural sciences have emerged and developed.

Science

One of the most significant periods in all human history, but perhaps the least widely recognized, was that around 500 BC (800–200 BC) when, in the three distinct and culturally disconnected areas of China, India and the West, there was a genuine expansion of human consciousness: in China, Confucius, Laotse, and the rise of all the main schools of Chinese philosophy; in India, the Upanishads and Buddha, Zarathustra in Iran; in Palestine, the Hebrew prophets; and, in Greece, we witness at this turning point of human history, the appearance of the literature of Homer, the pre-Socratic philosophers, followed by the whole of that great legacy of Greece to human culture.

In Ionia, the Greek colonists established a vigorous and hardworking culture, flexible and open to many influences. It was a time of travel, movement of populations, breakdown of the old and rising of the new. It was in this milieu of fluidity and change that science was born. The earliest scientific documents that we possess that are in any degree complete are in the Greek language and were composed about 500 BC. The Ionian Greeks brought to bear in their questions about the natural world a systematic and rational reflection that was distinctive and has remained the central characteristic of science ever since.

We find Thales (b. 625 BC) asking the question "What is everything made of?" He is the first person we know to look behind the infinite variety of nature for some single principle to which it could be reduced and be made intelligible. It is significant that in this search for unity behind the diversity of things the Ionians refrained from evoking any of the deities and mythologies of nature that were to be found in Homer and Hesiod.

Later, having moved westwards, the Pythagoreans discovered the significance of numbers but they were handicapped by the want of adequate instruments of research. Although they thought it vulgar to employ science for practical purposes, they gave us brilliant anticipations of modern discoveries. Yet, as Sir Richard Livingstone said,

> [Their] real achievement … was in fact that they wanted to discover and that by some instinct they knew the way to set about it … they started science on the right lines … the desire to know … the determination to find a rational explanation for phenomena ….[with] open-mindedness and candor … industry and observation.

(Livingstone 1923: 414)

So science was born among the Greeks. But with the coming of Roman dominance, although science continued, somehow its flame flickers to only a dim glow. From here the torch is, as it were, handed on to the Muslim culture. We in the West often forget that Muslim science lasted for nearly six centuries – longer than modern science itself has existed! When one becomes aware of the full extent of Muslim experimenting, thinking, and writing (in Arabic), one sees that without the Muslims European science could not have developed when it did. They were no mere transmitters of Greek thought, for they both kept alive the disciplines they had been taught and extended their range. When Europeans became seriously interested in the science and philosophy of their Saracen enemies around AD 1100, these disciplines were at their zenith; and the Europeans had to learn all they could from the Muslims before they could make further advances. Hence Islam was really the midwife to the Greek mother of our modern, Western scientific outlook.

The reception in the West of Muslim science and philosophy and that of the Greeks laid the foundation both of medieval natural philosophy and of that remarkable awakening in the sixteenth and seventeenth centuries to the power of human reason, especially in the form of mathematics when combined with experiment, to interpret natural phenomena. It is well authenticated that those involved in this development saw their activities as an outward expression of their Christian belief both in the orderliness of a world given existence by a God who transcended and instantiated human rationality; and a belief that the world, as the free act of the Creator, was contingent, so that how rationality was embedded in it had to be discovered by experiment. Kepler and Newton regarded the enterprise as "thinking God's thoughts after him." So a monotheistic culture was an intellectually appropriate matrix within which the natural sciences could flourish, even though it involves certain famous misunderstandings and adjustments in the relations between this new knowledge and that assumed by the traditions of the church and in the interpretations of Scripture. Historians have shown that the boundary between "religion" and "science" was always a very fluid one and differently located in different individuals, different societies and at different times.

From that origin in the West some four centuries ago has arisen the modern world in which science dominates our intellectual culture and, I believe, will continue to do so in spite of postmodernist misgivings. For the claim of the sciences to refer to and depict a natural reality other than ourselves is continuously and pragmatically vindicated by the successful technological applications of those same sciences. This is enough for most people to maintain its eminent position in any hierarchy of reliable knowledge. As an intellectual enterprise science is characterized by vigor, openness, flexibility, innovativeness, a welcoming of new insights and ideas, and a genuinely international, global community. In all of these respects, it stands in marked and usually favorable contrast to the public image of religious communities, including Christian ones. *They* tend to be seen as, if not lethargic and supine, yet as closed, inflexible, unenterprising, and immune to new insights, continually appealing to the past, to the "faith once delivered to the saints," and socially divisive as different Christian and other religious communities clash. So the Christian churches certainly have an uphill job to commend themselves globally to a humanity aware of the vast new vistas and opportunities now before it.

More particularly, there is, in the West at least, a collapse in the credibility of *all* religious beliefs as they are perceived as failing to meet the normal criteria of reasonableness: fit with the data, internal coherence, comprehensiveness, fruitfulness, and general cogency. Yet spiritual hunger is endemic in our times; attempted satisfaction leads to many aberrations in the so-called "new religions." Our society is, to my observation at least, full of wistful agnostics who would like to be convinced that there *is* indeed an Ultimate Reality to which they can relate.

Cultural transitions in Christianity

Religion in general has been defined as "a cultural sign-language which promises a gain in life by corresponding to an ultimate reality" (Theissen 1999: 2). Through its language, symbols, rituals, scriptures, music and architectural "sign-language," the Christian faith has promised the fruition of human existence in profound and eternal relation to the Ultimate Reality of God as manifested and made effectual in and by the life, death and resurrection of a particular person, Jesus of Nazareth.

More than almost any other religion, Christianity has elaborated a complex conceptual system of beliefs to give intellectual coherence to its intuitions and practices. What is affirmed, how it is affirmed, and what sort of metaphors are utilized to elaborate its system of beliefs have, much more than most Christians would admit, continually changed – and sometimes with an abruptness comparable to that of the paradigm shifts said to characterize the history of science.

In the two millennia of Christian history one can identify many such transitions induced by the facing of threatening challenges which generated within Christianity a new sense of vitality and relevance. In the very earliest days, recorded within the pages of the New Testament, one witnesses Paul's struggle to take the insights of the first Jewish followers of Jesus – claimed to be the hoped-for Messiah, the "Anointed One" in their terms – into the wider Jewish diaspora (hence his analyses of "law" and "grace"). Even more daringly, Paul deliberately, as the "Apostle to the Gentiles," entered the wider Hellenistic culture. His journeys from Jerusalem to Athens and then to Rome symbolize a profound transition in and challenge to the faith and experience of the early Jewish witnesses which was magnificently expanded, enabling Christianity to become the conduit, some two-and-a-half centuries later, of the religious impulses of the whole Roman Empire.

It then had to come to terms publicly with the intellectual life of that empire expressed as it was in the sophisticated and philosophical terms of a modulated Hellenism. It was to this challenge that the Cappadocian Fathers (Gregory of Nyssa, Gregory Nazianzen and Basil of Caesarea) rose when they articulated a system of Christian beliefs consistent with and in terms of the most convincing philosophy of their day. They out-thought their opponents both inside and outside the Christian church.

The arrival in the West during the eleventh and twelfth centuries, through the mediation of the Muslims, of great swaths of Greek literature, and notably the intellectually comprehensive works of Aristotle, constituted a potentially traumatic challenge to the received beliefs of Christendom. To this Albert the Great and his pupil Thomas Aquinas responded so effectively that the intensive and intellectually powerful synthesis of faith and reason of the latter dominated the church for more than six centuries. It is still today an intellectual construct that Christian philosophers ignore at their peril.

Apart from certain famous *contretemps*, the emergence of what is identifiably modern natural science in the seventeenth century was nurtured by its advocates and practitioners in a way that was seen by them both as consistent with and a natural consequence of their general understanding of nature as "creation" – that is, as being given existence by a transcendent, Ultimate Reality, named as "God" in English.

However, the following century too readily interpreted the astonishingly successful Newtonian science to imply a natural order that was so mechanistic and clock-like that God was often relegated to the role of the original Clock-winder. This concept of the inevitably absentee god of deism undermined the belief of Christians (and indeed of many adherents to the Hebrew Scriptures) in God as "living" and immanent in the processes of the world. Yet, in the nineteenth century, Darwin's discovery of the evolving nature of the biological world and of the role of natural selection, which entailed for some the final demise of a God no longer needed to account for biological design, actually reinstated the idea of God as *creating all the time* through natural evolution. As one Anglican theologian said in 1889, "Darwinism appeared, and, under the disguise of a foe, did the work of a friend." Nevertheless, the supposed "warfare" between science and religion imprinted itself on the popular mind in the English-speaking world, not least after the 1880s because of purely legendary accounts of the 1860 Oxford encounter between the Bishop of Oxford, Samuel Wilberforce, and Thomas Henry Huxley.

An uneasy truce between science and the Christian religion prevailed, each thereafter preserving a demarcated field for itself. It took over a hundred years, until the middle of the last century, for it to become apparent, with some notable exceptions, to a number of thoughtful scientists who were also Christian thinkers that the situation was not that simple and that the whole relation of science to religious belief, in particular to

Christian belief, was ripe for reappraisal. The existence, for example, of the Center for Theology and the Natural Sciences, which has made such a distinguished contribution to this reassessment, is – if, I may so put it – but the "tip of the iceberg" of a burgeoning plethora of activities in the field of science-and-religion. This has, up till now, taken place mainly in the academic world but, because of the prominence of ethical issues, is now spilling out into the life of the churches and that of the wider society. Present academic activities include: the development of an increasingly sophisticated literature; the establishment of academic centers in North America and Europe and of societies devoted to these issues; the publication of international journals in the field; the presentation of public lectures and a swarm of conferences and symposia; the funding of academic courses (greatly assisted by the Templeton Foundation); and – at long last – the beginning of funding of permanent academic posts in this field. All of which has led to …

The challenge to and reinvigoration of Christian thinking today by science

What characterizes science, as we saw, is a method which has been manifestly capable of producing reliable knowledge about the natural world, enough for prediction and control and for producing coherent, comprehensive conceptual interpretations of it. Such authority as the scientific community has can always be called into question, even though no individual scientist can ever repeat all past experiments, which have to be taken on trust. Yet, the scientific community has a limited and never absolute authority. The mere existence of such a method and of such a corpus of reliable knowledge resulting from it is in itself a challenge to traditional religious attitudes. I believe the time has come when mere assertion of authority by religious leaders and communities of the kind "The church says," "The Bible says," "The Magisterium affirms," will no longer carry conviction – for all such pronouncements are fundamentally flawed because they are circular and unable to justify themselves except by quoting themselves, or each other! Pronouncements of authority cannot be at the same time both self-warranting *and* convincing. Truths asserted in the promulgations of the ecclesiastical, scriptural or other authority cannot avoid running the gauntlet of those criteria of reasonableness we have already mentioned: fit with the data, internal coherence, comprehensiveness, fruitfulness and general cogency. Theology, like science, can claim to depict reality only if it is subject to these criteria and accepts that its formulations, couched inevitably (like those of science) in metaphorical language, are revisable in the light of new knowledge and perceptions.

To convince our contemporaries, the theologies of all religions must operate by inference to the best explanation, applying the above criteria. We need – as Hans Küng has argued – a theology that is truthful, free, critical and ecumenical, that is, a theology that deals with and integrates the realities of all that is discovered to constitute the world, and notably human beings. This reality has been largely unveiled by the natural sciences in forms never dreamt of by the founders of Christianity and those of other religions – forms that have a splendor and scope that no previous generation of human beings has witnessed. For we are now aware of what has been called the great "epic of evolution" – of the cosmos evolving and expanding by natural processes some twelve billion years ago from the hot Big Bang to the formation of the galaxies, stars and planets; to the emergence of sentient life on planet Earth; to the arrival of persons, to the advent of a Mozart, a Shakespeare, a Buddha, a Jesus of Nazareth – and you and me. The fragility of each individual human life is now set within this cosmic context and humanity sees itself both as a part of nature – we *Nature?* are stardust – and yet *apart* from nature, as we survey it from our subjectivity looking outwards.

Such a vista cannot but change how we view the physically based nature of humanity and the destiny of humanity in the divine purposes. At the same time it invigorates and enhances many strands of the received tradition, giving them a new significance in this wider vista. Certainly Christian hopes, for example, acquire a new pertinence but also a new context. Let us examine some instances.

God

How we are to regard God's relation to the world is challenged and enriched by this vista. God's immanence in the creative processes of the world now re-emerges with renewed cogency as an aspect of the divine nature, along with that transcendent otherness which all the Abrahamic religions must continue to affirm. God is all-the-time Creator. Creation is continuous, for God is creat*ing* in and through the processes of the world. God is indeed the "living God" of the Hebrew Scriptures.

This new emphasis on God working creatively in and through the very processes of the world, together with the recognition of the comprehensive explanatory power of the sciences in relation to those same natural processes, makes increasingly implausible any talk of God "intervening" in the world to change the course of events. God, as is said to be the case in the Christian sacraments, must now be regarded as operating "in, with and under" the world processes of which God is the circumambient Reality. For God is "the one *in* whom we live and move and have our being" (Acts 17: 28). This, for me, entails what I can only call a Christian pan-en-theism. God is the Circumambient, Infinite Reality in whom we live, by whom we are given existence and who works in and through us. We and the world are held in being and penetrated by God as a finite living sponge is held afloat and permeated by the endless sea (to use an image of Augustine). God works (instrumentally) in and through the processes of the world thereby effecting

God's purposes and ("symbolically") communicating Godself. (We could call this: "sacramental pan-en-theism.")

For me this insight involves taking absolutely seriously the affirmation in the Prologue to St. John's Gospel that it was God as outgoing, expressive Word (*Logos*) in creation which was all the time present incognito in the world, which was "made flesh" in Jesus – that is, made explicit and manifest in a human person. Such a recovered emphasis of ancient, revealed insights could both preserve Christian perceptions of the uniqueness of Jesus the Christ and, at the same time, recognize fully that God's Word, God's Self-expression, could also be manifest historically "at sundry times and in diverse places" (Hebrews 1:1) in other religions and cultures through their own symbolic and historical resources.

The stress on God's immanence at once also raises the issue of how we are now to conceive of God's interaction with the world and how God might influence some patterns of events to occur rather than others – as seems to be essential for understanding *inter alia* revelation and intercessory prayer. Scientific perspectives on divine action have been the focus of a series of state-of-the-art research consultations convened by the Vatican Observatory in conjunction with the Center for Theology and the Natural Sciences. Agreement still eludes these unique gatherings of theologians, scientists and philosophers, though the issues have been identified and, to some extent, clarified – but will, I am sure, continue to be on the agenda at the beginning of the new millennium.

God has traditionally been conceived of as transcending time and able to see past, present and future together, holistically, "all at once," as it were. Certainly space and time, since Einstein, are now to be seen as relations within the created order and given their existence by the Creator God. But whether or not God can logically *know* completely the content of a future that does not exist is hotly debated. The classical view is now widely called in question, and space and time seen from the new perspective of relativity theory are pertinent to the debate and profoundly enrich our understanding of God and eternity.

Biological evolution challenged received understandings of God's creative action when it showed that natural selection operates in living organisms whose form (phenotype) and functions have been modified by unconnected, random changes (mutations) occurring in its DNA. This role of chance led many Victorian thinkers to agnosticism. We, however, are now stimulated to see that chance and law together make not for an ossified universe, as in mechanistic pictures of the world, but one containing structures capable of change. God has to be conceived as creating through chance events operating in a law-like framework. This is a long way from the Artificer-Creator God, but perhaps nearer to a Composer-Creator God, weaving the fugue of evolving forms by exploring all the possible permutations of structure and processes inherent in the very stuff of the world, itself all the time being given existence by that same God. This generates significant new reflections on "natural evil," that is on the nature of pain, suffering and death – all of which are inevitable concomitants of an evolutionary, creative process that can elicit self-conscious, sensitive, aware *persons* capable of freely relating to the Creator God and co-operating in the work of creation.

Humanity

The evolved nature of human beings has generated particular problems for those who adhere to a literalistic interpretation of the Hebraic accounts of creation in Genesis in particular. (Is this a problem for Jews and Muslims too?) For human beings, who are a very late arrival on Earth compared with all other living organisms, never had, it now appears, a paradisal past but are rising beasts rather than fallen angels. Moreover, contrary to the implications of Genesis, individual biological death is the means of creative evolution and existed aeons before the appearance of humanity. Death certainly cannot be called, as St. Paul does, "the wages of sin" (Romans 6: 23) in any strict biological sense.

All of the foregoing, and much else, imposes on Western Christians, at least, the necessity to rethink those redemption theologies that are based on the postulate of an historical "Fall" (which was never actually propounded in the Hebrew Scriptures) and on Augustine's interpretation of "original sin." For if, as the scientific record now shows, human beings are creatures who have slowly, and sometimes painfully, emerged with self-consciousness, awareness of the values of truth, beauty and goodness, developed community and mutual co-operation – that is, "rising beasts" – what significance can we now give to the particularity of the "Christ-event," to the nature of Jesus of Nazareth? And what is he supposed to have done for all humanity? In what sense can we today affirm with the Nicene Creed that he "died *for* us" in a way that can be transformative here and now, and eternally, of the possibilities and potentialities of human existence? In all of this the Eastern Christian church's emphasis on the effect of the "Christ-event," the "work of Christ," as enabling humanity to be taken up into the life of God (*theosis*) is recovered and the profundity of Paul's emphasis on the significance of being "in Christ" is enhanced.

Furthermore, current advances in the neurosciences and cognitive sciences are showing how tightly linked the subjective mental processes which constitute our personhood and consciousness are to the physiological and biochemical processes of the human-brain-in-the-human-body. Christian anthropology has to return to the more Hebraic understanding of human beings as psychosomatic unities and not as the embodiment of naturally immortal souls – a notion imprinted on both academic and popular Christianity by centuries of the influence of Platonism.

These are profound questions. Many of us are searching for that rebirth of images which characterizes any truly vital community seeking a human relationship with God. Christians have some problems special to their received traditions, but because the problems have been generated by the comprehensive perspectives on the world and humanity the sciences now afford, they experience many of these challenges to received insights in company with the other great monotheistic Abrahamic religions. The world can now, with the aid of the sciences, be seen more convincingly than ever before as the creation of an ever-working ever-present Ultimate Reality, who transcends and yet is immanent in it – and can also be present in and to us humans. That is our hope, reinforced by our new perspectives on the cosmic process.

I believe that if religious believers of any faith ignore the new challenges of the dazzling, exalting even, vista which the sciences continuously amplify and spread before our eyes, we shall all – Christians, Jews and Muslims – simply be digging deeper and deeper holes. As we go downwards, we shall be talking more and more to each other and less and less to the great human world up there – a world now bathed in the clearer light of the sciences describing God's creative work. As we enter the millennium, light shed on creation from the sciences will transpire more and more to constitute the first glint on the horizon of a wider and deeper illumination of the cosmic significance of central, and sometimes forgotten, Christian affirmations which in this light become more and more reconcilable with those of the other Abrahamic religions and with our cosmic perspectives – but only if we are open to the enterprise at whatever cost to our pre-conceived notions. For after all, theology, like science, is a great enterprise of the human spirit.

References

Brooke, John H. (1991) *Science and Religion: Some Historical Perspectives*, Cambridge: Cambridge University Press.

Livingstone, Sir Richard (1923) *The Pageant of Greece*, Oxford: Clarendon Press, republished 1945.

Theissen, Gerd (1999) *A Theory of Primitive Christian Religion*, translated by John Bowden, London: SCM Press.

16 George Sudarshan

George Sudarshan is a Professor of Physics at the University of Texas at Austin. Prior to this he was a member of the Rochester and Syracuse University faculties. Sudarshan's main research interest is particle physics. In 1957, he co-discovered the Law of Weak Interactions, one of the four forces of nature. He also formulated the "Optical Equivalence Theorem" which established a bridge between classical coherence theory and quantum optics. His contributions cover a wide area of theoretical physics. His recent books include *Pauli and the Spin-Statistics Theorem* (World Scientific, 1997), co-authored with Ian Duck, and *Doubt and Uncertainty* (Perseus, 1998), co-authored with Tony Rothman.

Interview by Philip Clayton

PC: I would like to start by asking you about your religious background and affiliation.

GS: I was born in an Orthodox Christian family. I was very deeply immersed in it, and so by the age of seven I had read the entire Bible from Genesis to Revelation two or three times. I was not quite satisfied with Christianity, and gradually I got more and more involved with traditional Indian ideas. I would now say I am a Vedantin, with these two religious and cultural streams mixed together.

PC: Did your training as a scientist contribute at all to your growing dissatisfaction with the church?

GS: No. It was simply that I found that the people who professed to practice were really not practicing. In other words, there was a great deal of show and not that much genuine spiritual experience. Further, a God "out there" did not fully satisfy me.

PC: Was it a smooth movement for you into the Vedantic tradition?

GS: The earlier part was very smooth in the sense of becoming more like St. John than like St. Paul. But eventually when one had to make the statement "I am now accepting this and not the other." That was hard, because giving up anything is very difficult.

PC: You said that it was sort of a movement from Paul to John. Do you mean away from the legal practice and toward focus on theory?

GS: I mean from the prescriptive to the experiential. John talks in a very feeling fashion, while Paul is telling everybody else the right thing to do. John's way of looking at it makes it possible to be one with the universe, and one with God, and if you are one with the universe, then of course there is no question of not being a scientist. If you are one with God, there is no question of being irreligious.

PC: Suppose I had talked to you during your graduate training in physics, would you have tried to convince me that it was possible to be both a Christian and a scientist?

GS: I would have said it would be possible to be both. Probably not for a born again Christian – you couldn't be a Robertson. But you could be a Billy Graham; you could be kind to people and not deviate from your path of investigating the universe. Jesus said to Thomas, "My dear fellow, since you said you wanted to put your finger in the hole in my palm, come and do it." He was not against experimentation and direct personal verification.

PC: Do you see the religious and the scientific as just parts of a single quest for knowledge and truth, as complementary quests, or as radically different?

GS: I would say that they are parts of the same thing, but not themselves the same. Eugene Wigner said many years ago, "I don't think physics deals with everything. Whether I am happy or unhappy, whether I am afraid or fearless, whether I am noble or I am mean, how does this get represented in science? Even if there are people who would say there is a chemical imbalance, I would like to think that there is something else." I feel the same way.

PC: Some people, including many non-scientists, have the view that science controls the process of knowledge, whereas religion is a matter of accepting passively what is given to you. How does that strike you?

GS: I think that is an inadequate view and that it is propagated by people who have done neither the religious quest nor the scientific quest very seriously. Both in religion and science, the quest, it seems to me, involves doing everything that you can with the best of everything that you have. The immediate job of God is to make me see things which I could not see by myself.

PC: The first general area I wanted to ask you about was this religious belief that there is a purpose and destiny in the created universe. Christianity and Judaism and Islam say that not only is there an arrow of time, as physics might teach us, but there is also a *telos*, a goal toward which the universe is moving and for which it was designed. Do you see this religious notion of purpose as being compatible with the study of physics?

GS: I see them as different. In fact, my cultural background makes this one directional time and *telos* appear somewhat dubious. When I sleep, or when I dream, I have an entirely different time sense. Since much of my awareness is connected with those things, the statement that time always goes forward seems to me not very good. The equations of physics are symmetrical with respect to time, but if you disregard all the interconnected webs between things which are going backwards in time, keeping only the things going forwards in time, then you are tempted to say "See? All the sets are going forward in time."

PC: Do you think our understanding of God has grown through our advances in knowledge of the physical world, or been untouched by the advance of science? Or has the advance of science in some way challenged the traditional notion of God?

GS: The advances of physical science have made it unlikely that God is directly intervening with regard to the physical universe. The only time I would say that God would be directly involved is at those times when you have the highest experiences, when you feel that in fact it is not your doing but your witnessing. When that happens, you start wondering. You feel that there must be a subject to the action, and you attribute it to something outside of you.

PC: It sounds like you are saying, let's find the space for God's action elsewhere. What about the advance in the social sciences? If we had more powerful predictive theories in the social sciences would that take away this space for God?

GS: I don't think so. The so-called advances in social sciences seem to be more methods of manipulating societies – we are much better at persuading people to buy soap or candy than making people behave morally.

PC: What would you say to someone who told you they couldn't picture God as a transcendent being, but that they believe in the ethical things you are talking about – friendship and love – and it's fine if you want to call that spiritual, but the notion of God for them is too metaphysical. Could that person be religious without it?

GS: If a person feels allergic to the word God, but recognizes God in some other fashion, in terms of being kind or helpful, being self-sacrificing, and he is happy, then he has found God. She who delights in creativity, insight and harmony beyond her making, she has found God. If she finds herself with no unfulfilled wish, but is happy, then she has found God. If somebody needs organ music and cathedral ceilings for that purpose, good. If somebody else says they can do it while taking care of poor people, that is good too.

PC: Do you still find the notion of "God" to be helpful in understanding your religious life?

GS: I am a little divided in my thought. Normally the Western notion of God is semi-masculine, and God is above all things. God doesn't really need people, but people need God. I like a more feminine kind of God, a God who maybe identifies with grace. In the Vedanta systems, there is this branch called Vishista advaita which has the notion of "grace," referred to as *sri*. *Sri* is like the mother, and the mother's affection has no reason. It exists because you are her child.

PC: So it's very close to the Christian notion of grace as unmerited favor and affirmation?

GS: Rather than say unmerited, I would like to say a shower of blessing which does not count the cost, because some merited people also get grace.

PC: Have you integrated these religious insights into your work as a physicist?

GS: They contribute only in the following sense: I am a professor and I have been teaching for a long time. When I teach, I consider the communication of knowledge to be part of my obligation to my divinity. Similarly, when I learn from somebody, I consider that too to be a gift from the divine. Teaching and learning are in fact divine acts. In my own research I feel it is a task that is part of my spiritual exercise. But somebody who is watching it may say, "He looks just like everybody else."

George Sudarshan

One quest, one knowledge

Introduction

In the Hebrew Bible, there is a story about a remarkable man to whom
Abraham pays his tithe. His name is Melchizedek. Unlike most other people
whose genealogy is known and recited, Melchizedek has no clan, no
genealogy, no birthdate and no fixed address. As a priest, he is a figure who,
in a sense, resides outside time and space. To Hindu eyes, he represents the
eternal hidden within the temporal. In this essay, I would like to discuss the
relationship between the scientific and spiritual quests from a Melchi-
zedekian perspective or, as the Apostle Paul said to the sophisticated
Athenians with their many gods, "I want to talk to you about the unknown
God." I will focus more on the spiritual aspect of the Hindu perspective and
not at all on its institutional history because my tradition represents not so
much a particular religion as it does the human spirit; and "Hindu" is used
here as the old Greeks and Persians used it to denote the way of life of the
people of a geographic location.

Experiences and experiments are interpreted differently by different
people. Many find both the universe and their own lives to be magnificent,
well-designed creations. Others find only meaninglessness and purposelessness. When I was in my teens, I read William James's *The Varieties of
Religious Experience*. In addition to the religious experiences expressed so
beautifully by James, we need to add the contemporary anti-religious experience many people have of feeling indifferent toward reflecting and
meditating on their lives and the world around them. This is obviously not
entirely true – otherwise the SSQ conference would not have drawn such a
large audience – but it is a prevalent characteristic of our contemporary
culture. It seems to apply to many engaged in academic pursuits, and especially to those who work in physical sciences.

The scientific attitude

While there is a generally accepted framework for scientific research, no
unalterable dogma of scientific method exists. Science evolves and stands
always ready to be corrected in the light of new evidence. Although not

everything has to be experienced directly, all scientists, skilled researchers, are intimately and thoroughly acquainted with the broad issues of their discipline and can extrapolate from them to their own experiences. The scientific attitude is not unlike the attitude which Saint Thomas, one of Jesus' disciples, took toward Jesus' resurrection: "Unless I see the mark of the nails in his hands, and put my finger in the mark of the nails and my hand in his side, I will not believe" (John 20: 25). The basis of faith, there-fore, just as much as the basis of science, must be in things directly seen or experienced. Scientific research is, on the one hand, an examination of the orderliness and comprehensibility of the universe. On the other hand, it demands careful, critical examination of theories and experimental findings at every step. Not all scientists draw the same conclusion from the same data. The meaning, lessons and design of the physical universe are seen and evaluated differently by different well-informed, skilled researchers.

Whether a scientist thinks that the universe is designed or not, that it is meaningless or not, does not in general affect his or her scientific perfor-mance. But as soon as someone asserts that his views on design in the universe, or lack thereof, is the only scientific position, this view becomes a dogma. This is bad science, and it is bad *for* science. Similarly, differences of interpretation exist with regard to the world at large, the society in which we live, and the decisive events of our own lives. Whereas some see a benevolent agency guiding us along, others see nothing but chance and individual choice. Both views are possible.

However, if you are inclined to find meaning in the experiences of your life, then you will also be more inclined to find meaning within the world of science, too. This does not necessarily mean doing science any differently, but it opens the possibility of seeing science as a source of joy. In this case, science ceases to be an end in itself and becomes a spiritual discipline. In a Hindu hymn called "A Thousand Names of Vishnu," the very first phrase says, "The world is Vishnu." This hymn calls us to see the world as a manifestation of the Lord, just as the Psalms call us to glorify God for the magnificence of the creation. Scientific bigotry insists that there is only one way to understand science. One might suppose, for example, the path of science is separate from the path of the heart. While I would not wish to deny those who take this view the right to their opinion, I do hope they are wrong. Most importantly, all scientific knowledge, as a model of the world, is subject to rethinking in the light of new discoveries, both theoretical and experimental.

Purpose and discovery

In modern science, making a reference to teleology is considered bad form. In spite of this, however, even scientists plan their lives with some good in mind. When we look back on our lives, even random events appear to have been part of a plan. We can see looking back that there was indeed a pattern to the movement of our lives. In fact, the best explanation of your life, if you only knew the history of yourself in advance, is in terms of teleology, that is, in terms of some greater purpose. The most adequate explanation of everything that happens in your life is that there is a design, and that this design is being carried out.

I believe this to be true, even in cases where apparently bad things happen to good people. When I was an undergraduate, I became very ill one year around the time of my annual examinations, so ill, in fact, that I could not appear for my exams. I was put in a college hospital where the doctor very rarely came to check up on me. As I did not speak the local language well, I could not converse with the infirmary attendants. I had a lot of time to reflect on myself and my life. To add to the isolation, two students from a nearby high school who had chicken pox were put right next to me. Thus, anybody who might otherwise have come to see me, did not. At the time I thought, "My God, what is the value of this? Why did this happen to me?" And yet, in looking back, I can see that this time was the moment when I started studying physics very seriously. I was fortunate to have the time to reflect on my life, physics and the whole universe.

The openness of scientists in the moment of discovery is one of impersonal knowledge manifesting itself within, rather than one of discovering something outside oneself. This particular point deserves emphasis because many scientists are very careful to avoid any talk about the role of their personal experience in their discoveries. In general, people try to appear civilized and not talk about how personal experience figures into their scientific accomplishments. In the Hindu tradition, however, personal experience is the ultimate authority with regard to all things. All scriptures, all sayings, all teachings, are only a guide for organizing one's life. If things don't work out on a particular path, find another. If this works, then you have found what you were looking for.

Within my tradition, much emphasis is placed on the moment of discovery. Such insights need not be earth-shaking. They could be something quite trivial or small, but nonetheless they involve discovery. And when you discover something new, several powerful things happen. One is that you experience great joy. At that particular moment, there is nothing wanting, nothing you don't possess. Second, the discovery is something radically new. Humans have not been here before, but now this piece of knowledge has been uncovered for the first time. Finally, and somewhat paradoxically, at this very moment, you also see that this discovery is *not* something new, but in fact is something very familiar. There is no sense of strangeness in discovery, only a sense of belonging. It is as if you were returning to your point of departure. The cyclical nature of the world, as well as what you are in your own true nature, is uncovered in the moment of discovery.

Where is God?

Many people believe that the universe has a purpose, and that it is guided by a benevolent, sustaining agency. The West calls this agency "God," and asserts that we creatures belong to a powerful Creator who is far greater than anything we can conceive in our awareness or insight. Thus, within all the Abrahamic faiths, one must approach God with fear and trembling. To identify oneself with God is a blasphemous act worthy of banishment from society. In the past, drastic punishments were inflicted upon persons foolish enough to claim to be God, but these days civil laws protect them from such consequences.

The vibrant spiritual tradition belonging to my part of the world, namely the Hinduism of central and south Asia, believes instead that God manifests Himself, or Herself, in many ways and in many contexts. My tradition affirms that any spiritual search, whether academic or not, is bound to lead to God. Within Hinduism, there is nothing which is not sacred. God is not an isolated event, something separate from the universe. God is the universe.

Yes, God is more than the universe, but He is the universe. Therefore, there is no time which is not a time for prayer, no place which is not sacred, no event in which God is not present and involved. Consequently, someone from my tradition feels no sense of awe when encountering the divine, at least not in the sense of being filled with fear and trembling. How can you fear something which is yourself? This may seem to be a rather blasphemous attitude, to identifying oneself with God. At the least, isn't it an arrogant and egotistical misunderstanding of the true nature of God? Not if we examine the situation more closely. Not if we scrutinize the kind of God we are talking about.

As children, we wanted our fathers to be powerful. ("My daddy can beat up your daddy.") As we grow into adults, however, we want our fathers to be wise. The same goes for our models of God. Upon first thought, we want God to be all powerful. As our relationship with God matures, we want a wise God who can walk alongside us. Obviously, you don't walk alongside someone before whom you feel you must fall down in fear and trembling. Rather, in such a relationship, you expect to be shown certain things which you hold most valuable in life. In particularly insightful moments, riches lose their importance; only the feeling that wealth is significant makes them valuable. What we yearn for in our walk with God is permanence, happiness, enlightenment. All these things, therefore, must be aspects of God. If our awareness grows toward the eternal and permanent, then this is what we are yearning for. If our essential nature is happiness, then we seek happiness in order to be most fully ourselves in this world. The search for happiness is the search for our true being; it is, therefore, a spiritual path.

Hunting and gathering

Amongst those who devote their professional lives to science, there are two different categories of people. First, there are the gatherers, the integrators of available knowledge, who are especially adept at assembling knowledge, digesting it, and presenting it in such a fashion that most of us can understand. When we see how deftly a master presents difficult and complicated concepts, we think to ourselves that we could do it too, if only we tried. The gatherers are the ones who create consensus. They are guided by all the things happening around them, and they seek to put everything in its place. But they have a vested interest in things as they are. This makes them hostile to radically new ideas and frameworks.

There is, however, a problem with this kind of approach to science. Just as those who built the Tower of Babel needed to be of one mind and one tongue to achieve their goal, the scientific gatherers of our day depend upon a common vision in order to build their tower reaching up to heaven. But how many aspects of the mind and of reality must be covered over and concealed in order to achieve such unanimity? How much of reality do we miss in pursuit of a common goal? What degree of hubris lies beneath such attempts? And what are the consequences when our towers fall?

Norbert Wiener (1961), in his book on cybernetics, talks about the problems of frequency drift with regard to electrical generators. When one connects many generators together, the result is that each one will pull the other into synchrony, so that, by and large, the frequency of the electricity they generate in tandem does not drift. Wiener points out, however, that this doesn't really solve the problem, because when instability does set in, it is much wilder than before. The scientific gatherers of our day create systems that look impressive for all their clarity and scope, but they remain susceptible even to small disturbances in our knowledge of reality.

Then there are the hunters, those who go after knowledge. Hunters are not particularly sociable types. They don't seek agreement with conventional wisdom, and often devote themselves to one particular idea or experiment to the utmost of their ability. Often, such scientists end up traveling many years of their lives down blind alleys, with everyone feeling sorry for them in the end. Occasionally, though, hunters come up with a landmark discovery which is eventually integrated into the larger store of human knowledge.

In a similar way, the spiritual quest also must have hunters and gatherers. By and large, most people in religion, just as in science, are gatherers of wisdom. Not everyone can hunt, otherwise there would be few left to integrate the discoveries and insights, and what we had found would lack coherence. No one would be able to interpret the significance of the hunters' trophies. In a sense, religion as a whole is the gathering place of all spiritual insights and findings, whereas the individual's spiritual quest, if one is courageous, is the life of a hunter.

In my own life, I have been privileged to experience the joy and ecstasy of discovery in both the scientific and spiritual domains. In such moments, the distinction between scientific and spiritual paths vanishes for me. In fact, the feeling is identical for both. The majesty of the external universe, the power of the mathematical description of the world, the infinite creativity and insight one finds within, all of these fill me with a sense of joy, thankfulness and humility: joy, because it turns out that I am right, that I have discovered something true about reality; thankfulness, for the preparation I received which allowed me to receive this insight at this particular time; and humility, for having been entrusted with this vision.

Hinduism does not consider aesthetic aspects of reality to be different from the Godhead. True beauty is forever new. The divinity and purpose in the universe, as well as in our awareness, grasps and shapes beauty; it searches for beauty and meaning. Beauty is truth, and truth is beauty. Reality, therefore, is not something that can be grasped from only one perspective. A three-dimensional object can be recognized as such only by seeing multiple projections from different perspectives, with, so to speak, different attitudes, and by integrating them into one reality. They only appear contradictory to one another. For example, looking at a person face to face yields one aspect of that person. Looking at them sideways brings another aspect into view. These two views appear contradictory, and they *are* contradictory as long as they remain two-dimensional in our minds. But if we consider them in three dimensions, we discover that they imply not contradiction but depth, and can be integrated with one another, removing the contradiction.

In the Hindu tradition, then, the spiritual quest is in fact not distinct from the scientific, aesthetic or, for that matter, any academic pursuit. Once a year every year, we remind ourselves about the beauties of humanity by reciting certain verses which focus our minds upon the glory of human inspiration and insight. But we also recite verses having to do with each of the major disciplines: mathematics, astronomy, rhetoric, and poetry. These verses remind us that our sacred duty is not only to so-called "holy" knowledge, but to all knowledge because, in fact, all knowledge is holy.

References

James, William (1958) *The Varieties of Religious Experience*, New York: Penguin Books.

Wiener, Norbert (1961) *Cybernetics: or, Control and Communication in the Animal and the Machine*, New York: MIT Press.

Some concluding reflections

Philip Clayton

What are readers to make of the journey they have just completed? There is no question that we have traversed a rich variety of landscapes. One's first impression is of a rich, almost startling diversity; clearly the sixteen scientists who have spoken and written in these pages did not reflect a "party line." Instead, they represent a vast array of positions running from strict atheism through agnosticism to the various forms of devotion to God. Some draw very tight lines of connection between their science and their religious belief; others are cautious about even the most limited of connections.

Here, at the end of our journey, an overview and concluding evaluation might be helpful, a sort of guidebook to the terrain that we have covered, attempting to make sense of the rich yet complicated set of options expressed in these interviews and essays. Perhaps the following comments will offer such an orientation – an interpretive perspective, one alongside other possible ones, on the wealth of material contained in this book.

Four general features of the discussion

There are complicated things to be said about this terrain of science and religion, or science and the spiritual quest; we return to some of the intricate and subtle dimensions of these scientists' writings below. But before we begin to focus on more detailed features of the geography, it is good to pause and remember again the general lay of the land. The best way to do that is to recall the common paths that many of the authors have walked in their interviews and essays above.

1   *Science has radically and permanently changed how humans live in the world.* Every reader of this book will already be aware of the changes – just think of the technologies employed in writing, producing, transporting, advertising and selling this one volume! But one sees the underlying causes even more clearly by reading the reflections of these leading scientists, since they work daily with the theories and

technologies that have transformed our understanding of ourselves, our planet and our cosmos.

Of course, there was a time when religious thinkers did their reflecting largely independently of science – either because they did not think of it as an important source of knowledge, or because religion and science were placed in two separate worlds with few if any lines of connection. But in our day that situation has changed. Science, together with the numerous technologies it has spawned, is now a dominant force in human society, in human thought, and in our everyday lives. Religion, many of these scientists affirm, must either enter into dialogue with science or be demoted to an increasingly marginalized role within our civilization.

2   *Science presents a challenge to theism*. The theists in this book do not sound discouraged or hopeless. Rather, they express enthusiasm for the ability of Judaism, Christianity and Islam to co-exist with the best of modern science. But many of them do issue a warning: it is necessary for believers to grapple with the conclusions, and the worldview, of the sciences today. Arthur Peacocke puts the call with particular clarity: "if religious believers of any faith ignore the new challenges of the dazzling, exalting even, vista which the sciences continuously amplify and spread before our eyes, we shall all – Christians, Jews and Muslims – simply be digging deeper and deeper holes. As we go downwards, we shall be talking more and more to each other and less and less to the great human world up there."

3   *Still, science is not all-sufficient*. Among those who are the key contributors to humanity's knowledge of the natural world, one might expect the belief that the powers of science are unlimited. But rather than offering the highest possible estimation of the potential of science, these scientists go out of their way to stress the limits of cosmology, physics, biology and computer science. Each conveys a keen sense of what science can and cannot do. Some stress the fallibility of human knowledge; others emphasize the importance of ethical commitments that lie outside of science; still others note questions of transcendence and human meaning, of the "before" and "after," that science cannot answer. It's interesting to discover specialists in fundamental physics, the history of the cosmos, and the evolution of life converging on this common sense of the limits of human knowledge. Many of them give these limits a religious interpretation, as Ken Kendler does: "While God may be mysterious and we unable to understand His ways, yet we still have the right to ask, and the duty to try to understand. The relationship between humans and God is not one-sided. ... It is not blasphemy to question and seek to understand, to expect and hope that the Judge of all the world will do justice." But no one seems to think any longer that the power of science is unlimited.

4 *The category of spirituality represents an important bridge between science and the religious traditions.* Certainly not all of this cross-section of famous scientists is religious. The evolutionary biologist Michael Ruse expresses a clear agnosticism, and the neuroscientist and computational theorist Michael Arbib describes himself as "a Jewish atheist." Although some are active in religious communities and practices, others stand further from organized religions. Yet for all of them, the notion of spiritual questions and a spiritual quest is helpful, and for many it is central. More on this in a moment.

5 *The "spiritual quest" takes many different forms today.* When "Science and the Spiritual Quest" was launched in 1996 the project leaders believed that the word "spiritual" could bridge the gap between the large number of participants from different religious backgrounds and traditions. It was less clear what content, if any, would be given to the word and whether it would serve as a genuine bridge between the different perspectives of the participants. What you have seen in these pages is a beautiful record of how the spiritual quest is significant in the lives of a diverse array of leading scientists.

Some major approaches to science and spirituality

I believe four categories best summarize the major approaches to our question that have emerged in these pages. For some scientists, "spiritual" expresses their religious beliefs and the practices of their tradition. Cyril Domb's spirituality is expressed through the intricate practices of Orthodox Jewish observance, for Robert Griffiths through Christian belief and practice, and for Bruno Guiderdoni through the life of obedience as a Muslim.

Another group of scientists gives content to the spiritual quest through the ethical convictions they hold and the actions that follow. Thus Brian Cantwell Smith was a co-founder of Computer Professionals for Social Responsibility; Michael Arbib gives central importance to "what had happened during the Holocaust"; and Mitch Marcus believes in the limitations of science because "the moral really remains beyond our ability to formally discuss." In each case what we might broadly describe as "moral questions" provide a significant entrée into the spiritual quest and even become defining moments of it.

For another group the bridge is constituted by philosophical questions. Thus Andrei Linde and Geoffrey Chew believe that the nature of consciousness in the physical world is a key to understanding how there is still a place for spirituality after the massive success of the physical sciences. Brian Cantwell Smith explores the recent emergence of "intentional sciences" and speaks of a new "age of significance" in answering the question of spirituality. These thinkers draw no sharp distinction between the philosophical quest and the spiritual quest.

A final group of thinkers comes to the question of spirituality from the level of experience – from their experience of doing science and from the experience of existing in the world. Thus George Sudarshan focuses on experiences such as "the openness of scientists in the moment of discovery." He bemoans the fact that so many scientists "avoid any talk about the role of their personal experience in their discoveries." In his tradition, Hinduism, "there is no time which is not a time for prayer, no place which is not sacred, no event in which God is not present and involved." Jocelyn Bell Burnell puts the point with equal clarity: "Quakerism appeals to many scientists, for the openness and search that this experiential attitude implies is similar to the experimental approach of the research scientist." This is why she can say in her interview that she sees "strong parallels between the ways scientific understanding grows and the way one's knowledge of God grows." For these scientists, the key elements of the spiritual life are not religious doctrine and metaphysics, but rather an intense awareness of and attention to the richness of our daily experience in the world.

How strong is the connection?

The authors disagree on how strong the connection is between science and religion. None of them treats science and spirituality as existing in complete isolation. Indeed, presumably it is impossible for them to be *completely* isolated as long as the two dimensions co-exist within the unity of the individual scientist! Still, a range or spectrum of positions is represented in this collection. At one end of the spectrum, Allan Sandage is attracted to the work of the famous Danish philosopher Søren Kierkegaard, who wrote that "truth is subjectivity" and has nothing to do with matters of objective fact. As Sandage says in his interview, "I don't think there is any evidence. It is faith, not reason. If there is proof, faith is not needed!" He writes later, "Science and theology, to me, are completely separate, except to solve the mystery of existence." Yet even Sandage speaks of science and religion as "separate closets in the same house," noting that "both science and religion proceed by making models and testing them, discarding those parts of the models that do not work and refining those parts that do."

Most of the scientists lie along the middle of the spectrum. They observe some connections between science and religion, although the two still remain distinct. For Arthur Peacocke, the sciences "change how we view the physically based nature of humanity and the destiny of humanity in the divine purposes." Yet the connections also contribute to the theological task: "At the same time [science] invigorates and enhances many strands of the received tradition, giving them a new significance in this wider vista. Certainly Christian hopes, for example, acquire a new pertinence but also a new context."

Finally, for a few of the scientists there is no final distinction between scientific and religious or spiritual activity at all. The point is put perhaps most strongly and beautifully by George Sudarshan, "In my own life, I have been privileged to experience the joy and the ecstasy of discovery in both the scientific and spiritual domains. In such moments, the distinction between scientific and spiritual paths vanishes for me. In fact, the feeling is identical for both." And he concludes his essay with the words: "These verses remind us that our sacred duty is not only to so-called 'holy' knowledge, but to all knowledge because, in fact, all knowledge is holy."

Clearly, then, this is not a homogenous collection of scientists, interviews and essays. Some authors believe in God, others do not; some defend the possibility of miracles, while others vehemently deny that a miracle could occur. Some lend weight to the scriptures of Judaism, Christianity and Islam, while others do not. Even where there is agreement, the authors often follow different paths to get to the same destination. While the editors might have worked to produce a more homogenous collection, it would have lacked the richness of this more accurate document of our times. In its present form, this collection of viewpoints has the virtue of being honest. It is a snapshot of the debate on religion and spirituality today, a cross-section of the viewpoints and attitudes held by the leading scientists of our day. Many paths are opened up here, and the reader may find him/herself deeply drawn to some, and perhaps troubled by others. To paraphrase the famous expression: yes, there is an important link between science and the spiritual quest, and they are many.

A new way of being religious?

It is significant that not one of the authors characterizes the spirituality of his or her tradition as being in competition with the spirituality of other traditions. Instead, much more common is the stress on complementary insights. During the SSQ conference the Oxford glycobiologist Pauline Rudd, another one of the SSQ scientists, phrased the point beautifully: "We have a need to make tracks through the vast network of our experience. Multiple classifications of the world are possible. And who can say what is the length of the coastline of England?" For these scientists, it's clear that multiple perspectives on the spiritual world are possible. Indeed, many of the speakers at the conference stressed the fundamental unity of spiritual insight. As George Sudarshan commented, it's always the same sun, wherever you observe it.

Perhaps for this reason, the Science and the Spiritual Quest authors have generally placed less emphasis on *religious doctrine* than you might have expected. They have not spoken a lot about doctrinal purity, or about heresy, or about the propositions basic to their tradition. It's not that they don't disagree; they clearly have differences about whether God exists, about the nature of God, salvation, religious communities and many other questions. And yet one senses in many of the essays and interviews a willingness to rethink and reinterpret religious teachings in new ways as new information becomes available. Most of the authors may practice within a particular religious tradition and share the beliefs of their tradition. And yet it is rare to witness discussions between Jews, Christians and Muslims where there is so little sense of tension and conflict as in this volume. In light of the events of and after September 11, the importance of this common ground cannot be stressed enough.

Indeed, could it be that what is modeled here actually represents a new way of being religious? Clearly, for these scientists, being religious can never be pursued at the expense of high-quality science; it can never mean ignoring the best of what humans know about the natural world. And yet commitment to excellence in science is not (for these thinkers) inconsistent with commitment to religion – even commitment to a highly specific religious tradition with all its beliefs and practices. One wonders: could it be that these men and women, because they are scientists, are able to be religious in a different way than what one generally finds among religious people in society at large? Could it be that believers who are scientists show more of an openness to new experience, more of an acknowledgment of the diversity of religious belief, more of a concentration on what the spiritual traditions share in common?

This leads to my final question: Could this be, then, why the project from which this book has sprung was called "science and the spiritual quest," rather than "science and religion"? There is something broader, something more all-encompassing about the term "spirituality." As George Ellis said at the conference, "It is the breadth of the whole vision that is the center of spirituality: partial visions must be recognized as such." Many of those who spoke could not be called "religious"; indeed, some explicitly reject that label. And yet their broader scientific projects clearly represent forms of spiritual quest. Andrei Linde draws back from talk of God; yet he speaks of his quest to understand consciousness and draws parallels with the spiritual philosophies of India. And Brian Cantwell Smith hesitates to talk of transcendence; yet he offers an alternative to materialism, a world that is innately significant, pervaded with meaning.

Conclusion

Both science and religion begin from their own domain of experience. Both take their particular domain with utmost seriousness; they know what it means to engage in a rigorous study of the natural world. But when you raise questions of ultimate meaning, when you ask about the beauty of a mountain lake or a Mozart symphony, when you experience wonder at the sheer size of the universe or the complexity of its structures, you concentrate on experiences that are not, at least not directly, part of the subject matter of the natural sciences.

It is as if these scientific believers are saying to us, "Concentrate carefully on the world around you. Observe it. Seek to understand the best, the most predictive, theories humans have been able to produce. This is not an easy task, and it demands your best effort and attention. But such empirical knowledge is not all! For you must *also* lift your eyes higher. There are other questions for you to consider, questions that do not lie within the domain of the natural sciences. Yes, we are biological organisms; but we are also a species that is troubled by "the sense of something far more deeply interfused," by "a motion and a spirit, that impels all thinking things" (Wordsworth).

Perhaps a metaphor would be helpful. You might think of science as a sort of offshore island, as Professor Bell Burnell has suggested. Some of us, approaching it by sea, find ourselves confronted by steep cliffs: mathematical physics, the formulas of quantum mechanics, Hilbert spaces and Lorentz transformations. (You don't need the towering pine of Grand Unified Theory to find its heights staggering; a few years of college physics will do!) Others of us land more comfortably on gently sloping beaches – perhaps the study of ecosystems, or popularized versions of the results of the human genome project, or studies in social sciences such as cultural anthropology or the psychology of the family.

Whichever of the scientific fields you traverse on this island, if you know them well, you know that they are not identical with the fields and slopes of religious belief or spiritual practice. If you are a professional scientist, you will draw whatever connections you can between scientific fields and religious territories based on your intricate knowledge of both territories – or perhaps you will declare the two land masses unbridgeable, as some (Allan Sandage, for example) have done.

But what are those of us to do who lack the professional's intricate knowledge of the highways and byways of scientific practice? What about those who feel their religious practice and belief challenged by the results of science and the scientific mind-set? This group of readers may have derived some inspiration from these pages. Of course, unanswered questions remain, and deep conceptual problems demand further attention. Still, it is encouraging to listen in on leading figures in science and to hear them placing a deep stress on the spiritual life. Somehow, one feels, their success *on a personal level* at integrating the two shores of science and spirituality gives *us* permission to do the same. If some of the leading scientists, those "on the inside," can integrate these two dimensions in their lives, then we are somehow bolstered in our sense that integration may indeed be possible.

As this book appears, a second, even more ambitious project of world-wide dialogue on science and spirituality is in progress. "Science and the Spiritual Quest II" adds to the Western monotheisms the religious traditions of the East as well as the traditions of animism, paganism, nature spirituality and humanistic spirituality. As this new project progresses we are again finding that many scientists are passionately interested in pursuing discussions of this kind, formulating their positions and presenting their conclusions to the general public. Major conferences at Harvard University, Paris, Jerusalem, Bangalore and Tokyo are being planned to bring these scientists and their conclusions to the public eye. It is as if a vast wave of interest in the overlap regions of science and the spiritual quest is growing, overturning past prejudices and animosities as it races forward. Today it's too early to say what will be the outcome of these discussions, conferences and publications. One has the impression, however, that the scientific establishment – and modern culture at large – will not be left the same after it has passed. It appears that new tools and a new mind-set, emerging out of the world's spiritual traditions, are being brought to bear: to guide the application of science, to guide the development and use of its technologies, to put a more human face on scientific experimentation and application. One begins to hope that this new alliance will return our attention to a concern for the well-being of humans, animals and the earth's environment. Clearly it's a powerful alliance, perhaps the most powerful one in human history. One can't help but think it is a good thing – for science, for religion, and for the earth.

Of course, there are many battles still to be fought and many doubts still to be addressed. Perhaps much will have been abandoned, and much transformed, within science and the spiritual traditions before the day is done. But, for the moment, one reflects back on the testimonies that have been offered in these pages and feels some encouragement. Religious traditions are talking to one another; science is not foreign to the spiritual quest but, at least for these scientists, is in some sense integral to it. For a moment the sounds of battle are silenced and there are signs of a truce. Perhaps there is hope after all.

Index